8/11

Electric Bicycle Conversion Kit Installation - Made Easy

A Quick Guide on How to Define Your Usage Needs, Choose, Install, and Use an e-Bike Conversion Kit

Claude Rosay, MBA

www.eBikeBook.com

AR Publishing Company
Los Angeles, California
www.arpublish.com

Orders, inquiries, and contact info:
www.ebikebook.com
www.arpublish.com
www.ebikepool.com
www.ebiketouring.com

ISBN-13: 978-0-9800361-4-5
ISBN-10: 0-9800361-4-3

Library of Congress Control Number: 2010917655
Library of Congress subject headings:
Rosay, Claude
Electric Machinery – Transportation
Bicycle Commuting – United States
Bicycle Touring – United States
Cycling – United States
Electric apparatus and appliances - Transportation
Electric apparatus and appliances - Installation

Published and Printed in the United States of America

Riding a bicycle whether on street or dirt involves *inherent risks* that you the reader assume exclusively. Ride safely, and always wear a helmet and gloves! These types of electric bicycles are not for minors!!

To my family for their never ending love,
encouragement, and support

~ ~ ~

Contents

Acknowledgments

Thank you to my wonderful caring wife, Rosalina Rosay, who with her authoring and publishing experience, helped make this book a reality. Her book "Journey of Hope – Memoirs of a Mexican Girl" has been an inspiration to many and is currently used at many leading universities for everything from Child Development to Chicano Studies to Auto-Biography "how-to" college courses.

Thank you to Bob Cody (a good friend, Engineer, and Columnist) for his support throughout the Rosay Family's book writing experiences, and for his help editing this book. Thank you Bob, you're a class act!!

Thank you to my children Alex Rosay, Andrew Rosay and Ariana Rosay for being wonderful kids, and for their help in reviewing my work.

And last but not least, I want to thank the readers for their interest in a "green" technology that will add years to their lives, dollars to their pockets, and less pollution to the air that their families and my family breathe every minute of every day. You are early adopters not only of a new green transportation technology, but you are environmentalists. You should be commended by your family, friends, neighbors, and colleagues. Congratulations on considering this green transportation technology and I hope you find e-biking as inspiring as I have found it.

Authors' Notes

Most of the book was written to reflect my personal experience when I had to choose, install, AND USE my own electric bike conversion kit. This book is written from the perspective of my passion for cycling over the last 40 years and my environmental consciousness. I tend to pull from many personal cycling experiences. I hope you find my sharing of those personal experiences helpful in providing more context and insight into my opinions and advice.

I found that learning about this new technology took longer than needed and was fraught with the risks of making many bad decisions. In fact, I finally grew tired of trying to figure it all out and reach the level of understanding I am used to having when making technical decisions. So, I said to myself "Self, what the heck just go out a buy a good value, powerful, low end kit and then learn from that decision." And I did so gladly, if nothing else happy with the fact that an avid cyclist like me would not have to be perplexed to the point of inaction by the whole thing. I made a decision to gain experience by doing. So I bought my first kit fully not knowing if it would really meet my needs or not.

The info sources on e-bike conversion kits were too long, too expensive, too hard to find, too disjointed, too poorly translated, too poorly adapted from the international e-bike world, or too biased. Hence, I felt the need for a book like this written from the perspective of an e-bike conversion kit user and for the average person in North America.

As for me, I am just your typical spoiled, hard working Southern Californian native with an MBA from California State University of Long Beach. I have a 20 year career in software development for mid to large size companies. I wanted a book that spoke to me and my kind. I'm not afraid of tech, if it is relevant to my needs. No such book existed in 2010 to my knowledge that "spoke to my needs", so that prompted me to write this book.

I will tell you what I ultimately did in terms of the e-bike I built. Again, I used my cycling experience with my regular non-electric bikes in four styles to guide me. The four styles I own of regular bikes at road, hybrid, mountain and folding bike, like I said I'm a biking nut! Or a cycling enthusiast as some might call me. I started in 4th grade biking to school every day, a paper route by 6th grade, and I still bike every chance I can. In fact, most mornings I get on a stationary bike at home, while playing video games, sending emails, or reading during my morning miles. I used to own several dirt motorcycles, street motorcycles, and even an ATV but finally got smart before any serious injuries and sold all the motorcycles and ATV. Having three kids forced me to admit that "speed kills", and especially if that speed is on two wheels while dodging cars during the afternoon rush hour.

I hope you enjoy the following "quick-read" format without the mathematical equations, overly technical terms, and graphs, so you get to ride your new e-bike faster. In other words, just about any cyclist can get everything they need from this book to make a much more educated decision quicker and start benefiting from this new internationally popular Electric Bicycle technology.

●

It's important to note that this is a general guide to selection, installation, and usage ideas based on my experience. Riding a bicycle whether on street or dirt involves ***inherent risks*** that you the reader assume exclusively. As with most sports, significant injuries can occur, no warranties are implied or expressed, you the reader use the information contained in this book ***at your own risk***. Electric bike conversion kits should be used **only by adults, and not by minors.** We encourage only safe riding, and we ask you to always wear a helmet and gloves for your safety.

About the Cover

I have had a passion for digital photography for many years. Today, I have about 18,000 digital photos in my library. You guess it, I get to credit myself for all the photos and image formatting that are included in the book and covers.

The picture on the front cover is of a very popular and very inexpensive kit called the Wilderness Kit. It is the one I purchased and it is sold by hundreds of distributors. I bought mine through www.ebay.com (check my website for other sources of this popular kit and others).

The picture on the back cover is of my converted Diamondback Mountain Bike enjoying the sun and mid 70's temps on a weekend in November. This bike is my favorite form of transportation in the neighborhood. Of course assuming the weather and the scope of my "local errand" are permitting.

Introduction

I live in a glass house, so I cannot throw "green" stones, especially when it comes to the decisions that I have made about my automobiles. From my 1984 Toyota Pickup - 4 cylinder roots as a college student, I have evolved into a V-8 junkie although, my primary commuting vehicle is a V-8 vehicle and powered by compressed natural gas. With three teenagers, a Chevy Suburban was a necessity for shopping trips to Costco, coaching soccer/volleyball, going to the beach, and our beloved family road trips. I love our Suburban, but I have a new mode of transportation that I am passionate about and eliminated my exclusive dependence on a V-8 powered vehicle for many of my local errands.

I like to think of myself as an "environmentalist that went to business school." Meaning that I am pragmatic, cost conscious, recycle whenever possible, and do my best to make environmentally friendly decisions. My ultimate scenario is one where I can save my family money, and do my little part to "save the planet" at the same time.

In fact, it is this mantra that drove my wife and I to purchase a solar power system for our house that generates 6+ kilowatts per hour at peak performance. We literally got over 90% of our annual electricity in the past 12 months from the sun. Our house is still hooked to the neighborhood electric grid for off-peak/nighttime hours. So during the day we generate much more than we use and the electric company pays us for generating at peak rate, then charges us less at night off-peak hours usage. This off-peak electricity pricing incentivizes us all to do the laundry and the dishwasher at night as well.

My goal is to be more electricity conscious (i.e. turn off lights when leaving a room) and get to 100% solar powered in our house. The last 3-4 months we have been pretty close 95% of our power generated by our solar panels. The net effect is that the 25 year panels pay for themselves in 6 years. Though to be fair to the discussion on solar, the 6 year "return on investment" (not counting likely additional equity in the house) was calculated after 75% of our solar system was paid for by the local power agency and the Federal Government (45% and 30%, respectively).

Since system installation (about a 14 months ago), we have reduced my families carbon footprint on the environment by over 25 thousand pounds. That's over 12.5 tons of carbon dioxide that is not emitted into the environment while my family is saving money! Everybody wins including my neighbors that can barely see the panels on our second story roof. The more solar panels that go into the neighborhood, the less the electric company has to charge all of its customers for building new power plants.

It is this "save money/save the planet" potential about electric bicycles that has so inspired me and I believe many others. I envy the European and Chinese cultures for embracing electric bicycles, perhaps out of necessity, but embracing it none the less. Anyone that has travelled around Europe or China recently I'm sure would agree. My growing "environmental consciousness" has inspired me not only to jump right in and build an e-bike, but also to immediately publish what I learned from the experience of building an e-bike. Saving money and helping the environment are worthy causes, but of course I also really enjoy riding a bike more then I enjoy driving a car, weather, time, and agenda permitting. Driving a car in the city is often boring compared to the feeling of the wind in your face.

This introduction and book seek to inspire, educate and empower the electric bike novice to take action now. You do not need to be an engineer to understand and make use of this technology or this book. So now, let's define a few modern day issues that electric bicycles address, at least in part. Let's state the main benefits in one paragraph, your "30 second elevator pitch" when your brother-in-law asks you "why ride an electric bike, instead of driving a car".

We will define the electric bicycle concept. Also, it is important to cite some of the current government regulations and technological limitations. And last but not least, let's discuss how every one of us can help build a community of e-bike enthusiasts at www.eBikeBook.com to promote electric bicycles as a new and exciting green transportation opportunity.

What are the problems and issues caused by using automobiles for 100% of our commutes to work, and 100% of our local errands (as is the typical scenario for the majority of Americans)?

The issue caused by using automobiles for 100% of our commutes to work, and 100% of our local errands center around non-replenishing fossil fuels. First of course, there is a dwindling supply of fossil fuels. Demand has driven the price high. It remains high, and as we saw a few years ago it can go as high as $5 per gallon in the US. We are still very lucky or spoiled as many Europeans feel. In Europe, $10 per gallon is *the norm* and goes even higher than that during supply/demand imbalances. Needless to say, electric bicycles are much more popular today in Europe than here in North America.

In my opinion the United States is lagging behind Europe and other "early adopter nations" of which there are many, due to the fact that we are spoiled with cheap fossil fuels in comparison to much of the world. However, during the Winter Holiday Season of 2010, many large TV news agencies made compelling cases in my mind that we may see $5 per gallon or more in the United States again (in 2011 and/or 2012). A 50% or more increase in the price per gallon could and likely would be a significant factor in greatly increasing demand and usage of e-bikes in the United States in the next several years. So, there it is, "the glass is half full."

The automobile, while it has many admiral qualities, has and creates a number of problems and issues. First, it is a pollution machine that expels potentially fatal gases and pollution as soon as you turn on the ignition switch. I'll define the pollution issue caused by automobiles in the immediately following section.

Speaking of fatal, last year in the United States there were over 38,000 automobile fatalities in just one year! A staggering number that is 9 times larger than the loss of lives by our brave troops in Iraq since the war began years ago. I must repeat that the 38,000 deaths is an *ANNUAL* number of lives lost due to automobile accidents. Assuming a direct correlation between miles driven and number of annual lives lost a 10% annual reduction in annual miles driven nationally could save hundreds if not thousands of lives saved annually!

Automobiles are an expensive form of transportation that the majority of people in the world simply cannot afford. In the United States our culture has created a complete dependence on cars for everything from commuting to shopping to leisure. There is almost an unanimous expectation that every person needs a car and should use their car for every errand. My European born father was a chemical engineer and the ultimate pragmatist. He and my European family were always amazed by the waste of what my father called the "Automobile Culture" of America.

Looking at the personal costs of the vehicle, there is an "operational cost" per mile that does not include the "depreciation and insurance costs." The annual depreciation cost roughly is the lost value of the car if you put it in your garage and do not drive it for one year. Insurance costs would likely not be affected if you lowered you annual mileage by using an e-bike, but it is definitely possible especially for e-bike commuters. The operation costs are all the other costs like fuel, maintenance, brakes, tires, and any other wear and tear from use. Studies show the operational cost to be over 50 cents per mile for an automobile (compare this to 2 cents per mile of an e-bike and then this becomes an opportunity to make money very quickly). Basically, for every 100 miles that you ride your e-bike for an errand instead of drive a car, you will put $50 of cold hard cash back in your pocket. And I really like the feeling of multiple extra $50's in my pockets and the more the better!!

What is automobile pollution and how is it measured?

The answer is twofold. First is their impact on the global environment and global warming? Then, second we will review automotive exhaust chemicals and their disease causing effects.

It's impossible to dispute that our roads are filled by cars that exact a high toll on our society. The effect of automobile exhaust on global warming is cited in numerous scientific reports as the principle cause of global warming. Automobile exhaust is blamed for worldwide increases in carbon dioxide (CO_2), nitrous oxide (N_2O), methane, water vapor, chlorofluorocarbons (CFCs), and ozone. Highway vehicles (passenger cars, light trucks, and heavy duty trucks, and others) are estimated to constitute 77% of the total green house gas (GHG) emissions for the

United States. The superheating process used by catalytic converters to get rid of unburned fuel actually results in an increase in green house gas emissions (GHG) in the form of increased levels of carbon dioxide and nitrous oxide being expelled from the car's exhaust system. Again, 77% of these emissions in the U.S. come from automobiles!

I'll' go out on limb here, but I believe that Americans should drive automobile less, whenever possible. I feel that taxing fuels like they have done for decades in Europe may be the best answer to creating a cultural shift in the United States. For example, "local errands" are an excellent and easy way to eliminate several car trips a week. Do we Americans really need to drive a car to the bank, then come home and ride a stationary bicycle for exercise? The irony is obvious. It is important to know that *eighty percent* of the price of fuel in Europe is tax. (Mitigating Global Warming)

Air pollution is also a source of many materials that may enter the human bloodstream through the nose, mouth, skin, and digestive tract. A short list of the likely pathogens that are found in car exhaust:

- Carbon monoxide

- Nitrogen dioxide

- Sulphur dioxide

- PM-10 suspended particles (less than 10 microns in size)

- Benzene

- Formaldehyde

- Polycyclic hydrocarbons

These substances have been shown to produce harmful effects on the blood, bone marrow, spleen, and lymph nodes. Respiratory and nervous systems diseases are also of great concern from automotive exhaust. (Ref. Automotive Exhaust Chemicals: disease causing effects)

What are the benefits of electric bicycle technology?

The answer is more fully found throughout the book. From a high level, I believe an increase in physical exercise for the rider is personally important for each of us car culture zombies. If you eliminate 10% of your annual miles driven, then you eliminate 10% of your personal automobile exhaust, a very worthwhile goal. Saving money and lowering your annual operational car costs are easy to document and calculate. (See my website www.ebikebook.com for a cost savings calculator)

To sum this up, e-bikes offer the freedom and fun of cycling without the pain and heavy sweating. Not that sweating is bad, but do you really want to show up to work or the market in a full sweat? Recent innovations in new Lithium Ion battery technology, a worldwide focus on green transportation methods, and rising fuel costs have given the electric bike a new lease of life. For example in China where, according to a report by Austin

Ramzy in *Time*, June 14th, 2009, there were 21 million e-bikes sold in 2008, outstripping car sales in China by 11.5 million units. (Smeaton, June)

Definition of an electric bicycle (e-bike):

An e-bike is a pedal assisted, motorized bicycle that conforms to state and federal laws regulating electric bicycles.

An important goal for me to maintain is being free from vehicular registration and taxes. The "e-bike retailer mantra" is "The US Senate has passed SR 1156, clearing the way for a legal definition of an electric bicycle in the USA." President Bush has signed this law. The new law assigns the governance of electric bicycles to the Consumer Product Safety Commission, and defines a bicycle that has pedals, (and is capable of being propelled by those pedals) an electric motor of no more than 750 watts, and a top speed (on motor only) of 20 mph **_as a "bicycle."_** You may want to be able to recite the "SR 1156 e-bike law" to a police officer should you get "stopped" by an officer that calls into question your motorized vehicles' legitimacy and your right to use the roads or perhaps more importantly sidewalks, bike trails, or bike paths.

Limitations of e-bike technology:

The limitations to e-bike technology today will be greatly reduced as time marches on, of this I am certain.

The cost, weight, and size of current battery technology leaves a lot of room for improvement. The distance traveled is advertised as "up to 20 miles and results will vary." How can you plan ahead with that, or know exactly when it will run out? Though I am still yet to fully run out, I have gotten close to running out and the second half of the ride is in a low power mode. Recently, I have seen 48 volt systems claim "up to 75 miles" of distance in the $3,000 price range.

Or for $13,000+ currently, you could get into an Optibike boasting configurations with 100 mile ranges (see www.optibike.com). Optibikes are literally hand built in Boulder, Colorado. It truly is as Forbes called in October, 2010 – "The Ferrari of Electric Bikes". A mileage range of 50+ miles could make for an exciting "commuter" or "touring" e-bike needs would be well served. Although like most cyclist, I do not own a Ferrari even though I would like to have one. So the price point of these Ultimate Electric Bikes is a hindrance to wider market acceptance. The Optibike's 3 year warranty lithium battery is $1,500 to replace. (Optibike)

Ouch, seems like I could build two to four e-bike conversions for just the price of that replacement battery. I would be very afraid to lock up a bike like that outside a market or grocery store in LA. If you can swing that price, and can get good theft insurance, then you should consider buying this bike if you want the ultimate in electric bikes that are manufactured here in the United States.

The other spectrum of the prebuilt electric bicycle market surprised me recently. I saw on Amazon.com a complete prebuilt e-bike with the Currie motor and removable 24v battery system and 450 watt motor on a hybrid bicycle for ONLY $349 (includes the bike!!), wow I almost bought it for my kids! Luckily, I remembered that "e-bikes are not for kids" in my opinion. I wonder if my wife would use it. Seems like a great "recreational" starter e-bike in a flat area like (no hills). Maybe you could use it for light shopping or close commutes. Having said that, don't forget many people in the US cannot afford the $400 for purchasing this e-bike, or the conversion kit like the one I purchased and installed. So once again the price point for electric bikes, even the cheapest e-bikes or conversion kits, is currently a limitation of this technology.

Overall, the charge time requires a little planning and forethought, for my system it takes about 30 minutes for a full charge. Though, I believe the Lithium Ion batteries charge faster than my SLA battery. Road safety is another limitation to popular cultural acceptance and use. You risk your life every time you share even a minor road with a moving automobile driven by a person of unknown driving ability, health, mental state, distractions, intoxication, and age.

Building a community of e-bike enthusiasts:
It is my hope that www.ebikebook.com helps build a community of e-bike enthusiasts to promote electric

bicycles as a new and exciting green transportation opportunity. I imagine it will contain links to "fan pages" powered by www.facebook.com and that users like yourself will share pictures, ideas, and results of their e-bikes. It should become an on-line resource power by the community for the greater good of the community. At www.ebikebook.com there will be no membership fees or costs to participate. Not only is user participation and contributions encouraged at this site, it is intended that www.ebikebook.com (like www.facebook.com) will have the majority of content on the website be user generated.

Part One

Defining Usage Needs

Like any good decision making process that involves a technical solution, the better the problem and the environment are defined... the better the solution. This section is dedicated solely to help the reader define his or her needs. Having spent the last 20 years in sales for a software development firm, I have gained much insight into defining problems before ever thinking or talking solutions. At work we call this step the "Needs Analysis," a process that I do multiple times daily.

Without a comprehensive needs analysis, you may spend hundreds, if not thousands of dollars on a solution that at best you are not satisfied with and at worst does not work to meet your needs in any way. A good "needs analysis" as is present in this book will greatly reduce this risk and help you build a checklist of what you are shopping to buy. The options here are many and so are the chances to purchase the wrong kit.

Defining your usage needs

The needs analysis is focused on how you plan to primarily use the bike. A very important perspective can be gained from my personal experience. When I started on this project, I had a much broader and different view of how I would use the e-bike. I felt I would also use the bike for recreational use on bike trails. To date, I use the e-bike 100% for errands, and leave the Suburban in the garage as often as possible.

You'll be glad to know that every errand is now fun and that I actually look for errands to run so I can e-bike. In fact today, my wife told me that I didn't need to go to the farmer's market. I tried to find a reason to go, but she insisted it wasn't urgent and she would go later in the week. I was actually disappointed that I could not ride to the market. In contrast, have you ever felt disappointed about not being able to enjoy the drive in your car to the grocery store? I don't recall that I ever did. In other words, driving your car usually isn't as fun.

I still use my three remaining regular non-electric bikes (road, hybrid, and folding bikes) for recreational riding depending on my needs, riding buddies, and location. I have not used the e-bike for riding on the bike trails. I find myself doing less recreational riding now, hence less riding on my non e-bikes. I think because the e-bike is so much fun, even just for errands. I use it to replace my car for short urban errands (bank, groceries, redbox.com DVDs, ice, garage sales, visit friends, etc...). For example, if I know I can ride for 2-3 errands or more in a day, then I do not go out to the beach to ride my road bike.

Very importantly I now get to my errand destinations faster and less sweaty than when I rode a regular bike to do errands, as there are some rolling hills in my neighborhood. Also the ability to quickly ascend all hills is GREAT! I have an 18 speed mountain bike with three sprockets on the pedal crankshaft and six gears on the rear cassette. Eighteenth gear gets the bike moving the fastest of course. I leave my e-bike in gear 17 out of 18 gears the whole time. It's rare that I ever switch gears, only for the steepest of hills. I just change the front gear down to the middle sprocket for steep hills if I want to stay seated, so I stay in gear 11 out of 18 as my lowest gear. If the hill is extremely steep, I'll stand up on the pedals in 11th gear and make quick work of it. The "take-away" here is that you will get places faster on the e-bike, especially when loaded with groceries, etc.

The main point here is that you need to think about how you want to use an electric bicycle before you decide the very first three things, which are:

1. Should I purchase e-bike technology now?
2. Should I buy an integrated e-bike that is ready to use or a conversion kit to upgrade a regular bike?
3. What power and bike style should I use?

Let's look at these three questions, what do they all have in common? The answer is that it is impossible to answer any of them without first completing a detailed needs analysis to define your usage needs and wants. At the end of this Part One section we wrestle with answering these three questions, one at a time.

First, let's gather as much information as possible to define what you would like to accomplish with your e-bike technology. You can download a Usage Needs form from www.ebikebook.com to help you make a better informed and hopefully more appropriate decision.

Also, don't forget that a big part of the benefits of this exercise is to better focus your expectations prior to purchase, which will undoubtedly improve your satisfaction post purchase. Leave no stones unturned. Spending adequate time on the needs analysis will help you select the better or best fitting technology to satisfy your usage needs and expectations. And you'll benefit from more realistic expectations as this is a new technology.

You are an innovator and you likely do not have any peers to guide you in your e-bike project. I trust that you will feel like me when I did my homework before purchasing a conversion kit. I believe that as you process the information in this book that your expectations may change and even change significantly.

Here are the usage needs questions that you need to answer or at least consider prior to even deciding if e-bike technology is right for you. You can also download the Usage Needs Survey from www.ebikebook.com . Or you just photocopy these pages and begin to take notes. Or even just write in the book (in pencil, ideally). Every question is important, do not skip any. Answer the questions briefly, using bullets is fine. In Part Two I will review each question in detail and help the reader analyze their answers to gain insight into what type of e-bike and conversion kit will best fit there needs.

USAGE NEEDS SURVEY

GOALS:

What do you want to accomplish?

What is driving your interest in e-bike technology?

Do you want to save money by reducing the miles driven for local errands?

Do you want to lose weight and lower your cholesterol?

Do you want to improve your aerobic conditioning?

Do you want to set an example for other "automobile zombies" that it is possible to run errands on a bike without breaking a full sweat?

Do you have a budget for this e-bike project?

RIDER:

Are you an environmentalist by nature?

How many years have you been cycling?

Do you have experience riding a bike that is loaded with extra weight? (i.e. pulling a trailer or "tag along", newspaper route as kid, loaded baskets/rack, rode a friend that sat on the handle bars or rear rack, etc...)

Have you owned a motorcycle or moped? (Are proficient at driving a motorized two wheel vehicle?)

How much do you weigh?

Do you have lots of tools, and consider yourself handy with them?

Do you do some basic bike maintenance yourself?

Will you be the only rider?

If you not the only rider, then what is different about the other rider(s) in terms of skills, physicality, and needs?

Have you ever felt like MacGyver from the TV show (or Mac Gruber from SNL)? Do you want to be in a position where you might need to re-engineer some components or settings on your bike?

BIKES:

What other bikes do you have?
Do you have a mountain bike or hybrid that is not getting much recreational use?

Do you currently have a bike with a rear rack and/or baskets on it?

Do you currently have a bike that you use for local errands?

If yes, then how is that biking working today for local errands? Does the bike have the ability to manage the extra weight (about 40 pounds for battery and hub motor)? Do you hear "creaking noises" from the frame when you push down hard on a pedal?

Do you have a bike with suspension forks, shock absorbing seat, and/or rear suspension?

How many gears do you have on the bike you are targeting for the upgrade to an e-bike?

How strong is the frame on a scale of 1-10, with 10 being a new $800 mountain bike with front and rear suspension, and one being a 20+ year old rusty bike?

How old is the bike?

What condition is it in?

How are the brakes, tires, tubes, chain, gears, shifters, seat, grips and crank shaft?

What are the measurements of the wheel? (26", 700cm, 24", 20")

What is the width of the front forks at the axle line?

What is the width of the front forks at 5 inches above the axle line?

What is the width of the rear wheel mount at the axle line?

What is the width of the rear wheel mount at 5 inches above axle line?

RIDES:

Where do you plan to ride?

Will you be riding alone or with others?

If planning to use the e-bike with others, then do they have e-bikes too? (What kind and what power do they have? And how often will you be riding with others?

Are there safe bike lanes?

Will you be riding before dawn or after dusk?

Are there side streets, bike paths, and sidewalks that you can utilize for most of your ride?

How far is your average ride?

Are there hills?

Is there wind and how strong?

What is the maximum load of items like groceries that you want to carry? (weight and number of traditional brown paper grocery bags)

Do you want to ride to work?

How many miles to work?

How safe would the ride be for a 20-25 mph e-bike?

What is the traffic like at rush hour in the AM and PM?

What is the road condition? Pot holes? Gravel? Dirt?
Mud? Sand?

Have you spoken to your boss about you considering
riding an e-bike to work? Are you the boss?
What does management think about your e-bike parked
and charging in the warehouse or your
shipping/receiving area?

Do you plan on "Touring" with your e-bike?

ENVIRONMENTS:

Where will you store your e-bike? Indoors?

Is the bike storage location easily accessible? Any stairs?

Is it secure?

Can your kids or neighbors' kids get to it at one of your storage locations?

Do you plan to load the e-bike into a truck/van or onto a car bike rack?

What is the average weather like?

Do you plan to ever ride in the rain?

Will your bike sit idle for several months at a time?

Is there a place indoors to secure, and charge your bike at work?

Other Issues?

We will review each question above and analyze together typical answers to the Usage Needs Survey in depth in "Part Two – Choosing a Kit". *Ideally, do not skip ahead to Part Two. Just continue reading the remainder of Part One to gain additional perspective that should help you answer some of the questions.* Then, complete a quick review of your answers to the questions above now before you continue. This way your answers to the questions above are unbiased by the Needs Analysis information in Part Two.

Now let's consider the three main questions presented on page 20, that are at the core of your e-bike project decisions. You do not need to answer these questions yet as you should read Part Two prior to answering them for yourself. What I felt would be helpful right now is to give you information at a high level on how I approached these three questions.

The three questions again are:
1. Should I purchase e-bike technology now?
2. Should I buy an integrated e-bike that is ready to use, or a conversion kit to upgrade a bike?
3. What power and bike style should I use?

Let's review these three questions now.

Should I purchase e-bike technology now?
The answer is most likely yes, but there are many pitfalls and issues that would make it a firm NO. The only way to accurately be able to answer this question for you is to complete and give careful thought to your answers to the Usage Needs Survey. In fact don't hesitate to go back, review, and update your answers.

It's true that the electric bicycle technology and marketplace will go through an amazing amount of change and growth in the next 10 years, but in the next three to five years probably not so much. Look at the investment in an electric bike or e-bike conversion kit as a decision you should plan to live with and justify for the next three to five years. Hopefully the battery, motor, controller, and charger will all last three years for you, but that is far from guaranteed. So, that being said, consider replacement parts availability as you finalize your vendor and product selection.

More importantly, I contend that if you get an e-bike kit now, you don't have much to lose financially; in fact, it could and should even *make* you money! The Return on Investment (ROI) and payback on your investment in this technology is less than a year if you purchase a system like I did. This green technology has the potential to put big dollars back in your pocket.

In its simplest form, the equation is that you put 50 cents back in your pocket for every mile that you use your e-bike instead of your car!!! Heck the gas alone is 20 cents per mile or more for half the cars on the road. At 1,000 miles per year that is $500 back in your pocket. See our website at www.ebikebook.com for a spreadsheet that you can tailor to your needs, local gas prices, and your car's miles per gallon for a more accurate annual cost savings calculation.

Should I buy an integrated e-bike that is ready to use or a conversion kit that I need to install on my bike?

If money is not an issue, then you should consider spending $2,500 to $3,000 for a new integrated e-bike. For me, the initial purchase price point of a new integrated e-bike from an e-bike manufacturer, distributor, and service provider is the only reason I took on all the significant additional risks and hassles of dealing with a conversion kit. I'm glad I did, as now I will know a lot more when it comes time to make a much more significant investment, whether I build or buy next time, I will have much more experience.

As of early 2011, for a new, ready-to-use integrated e-bike with the power that you will be happiest with over time, plan to spend $2,500 to $3,000 for a state of the art new electric "pedal-assist throttle" bicycle. Look at the bikes from iZip, Trek, or from Ultra Sport. Those were my favorite manufacturers. Models with a 750 watt motor, and a 36 volt system were the ones that appealed to me the most. One of the things that the three companies do so well is that they hide the lithium batteries in the bike frame that reaches from the front forks ("gooseneck") down to the front sprocket and pedals. The e-bike technology seems to be evolving so fast and is still somewhat a new technology. Given that, I felt that spending $3,000 now seemed wasteful, since I cannot use it for commuting to work. I expect that prices will fall and hopefully in half at some point in the next 5-10 years.

Trek is one of my favorite bike manufacturers. I was very happy to see their entry into the $2,000 plus electric bike marketplace complete with regenerative braking that creates electricity to charge the battery when you apply the brakes. This is a very good thing since more stops, implies more starts, and starts from a dead stop are one of the biggest drains on the battery. Many hybrid automobiles don't even have regenerative brakes that charge the batteries. Unfortunately from what I could tell from the product brochure, it only has a weak 350 watt motor, much less potential then the 600 watt kit I installed on a 20 year old Diamondback mountain bike for under $350. To be fair to Trek, my favorite bike manufacturer, I have not tried their electric bicycle and do not really know how my 600 watt motor and power system stack up. I do know I have about $2,000 extra in my pocket. And that is a good thing!! I expect many good things from Trek in the next several years in regards to e-bikes. In fact I hope to buy one from them in about 5 years if I see a good fit and good value for my local errand needs.

The intent of this book is to focus on converting a bike that you own to a fast electric bike. We will not spend a lot of time talking about ready built e-bikes, that's what the vendors are in business to do. Even if you are going to get a new integrated e-bike soon, you may still want to start with a few hundred dollars worth of a conversion kit to get a feel for power, speed, and battery weight and distance. If nothing else, it will help you pick the best fitting e-bike out there for your long term needs. And at $3,000 a pop, who wants to make a wrong decision? Not me!! Hence, my decision was to install a conversion kit first. And get a feel for the power of a 600 watt brushless motor driven by 36 volts.

By 2015 or for sure by 2020, I will probably add to my collection a full suspension integrated e-bike with the latest in lithium ion battery technology. Hopefully, there will be more competition, lower prices, and better battery technology. All batteries (lithium ion and the old, heavy sealed lead acid batteries) are expensive and have a limited number of years of functionality before they need to be replaced. Lithium ion battery replacements today are in the $500-$600 range (and higher). Sealed lead acid (SLA) batteries weigh four times as much and are in the $200-$250 range for 36 volt environment.

Needless to say the lithium ion batteries are the way to go if you can afford them; I didn't use a lithium ion battery for my first bike. I've opted to build my new 20 to 25 mph e-bike in the most economical way possible. I'm doing so with the cheapest battery technology to keep the potential costs of battery replacements more affordable. This approach also allows me to worry less when I park my bike outside on one of my errands.

What power motor, battery, and bike style should I use for installing my conversion kit?

Most road bikes and most folding bicycles should not be converted to an e-bike. The reasons are many. For starters, those bikes cannot handle the extra weight very well and would be more difficult to control (less safe) than the alternatives. The only exception is a Touring Frame; basically a road bike engineered and beefed up to handle extra weight. Although on a touring bike, there is likely no suspension in the front, and definitely no suspension in the rear. There may be a shock in the seat post, but that's it. It would be a rough ride given the faster speeds, especially compared to the alternatives.

So what alternatives does that leave? It leaves the *Mountain Bike* as the strongest frame bikes with the best suspension if you plan to have baskets attached to the bike for *"local errands"*. Or if you are trying to build a fast and sleek *"commuter"* then upgrading your *Hybrid Bike* with its 700cm wheels would be a great starting point. The main idea is that bikes (especially e-bikes) are "purpose built". The Usage Needs Survey should ultimately point you towards the battery style, motor power, and even bike style that would be optimal for your needs.

Let's get down to specifics. If you want to go 25 mph on the flats with no wind on a mountain bike then you need a 600 watt motor. If you want to go 31 mph then you get a 1000 watt motor. (Slinn, 2010) The bigger the motor, the faster the battery drains. As you can easily tell by the 66% increase in watts used, and only a 25% increase in speed, the *law of diminishing returns* applies here too. (Slinn, 2010)

For more details and even mathematical equations on power, batteries, and distance start by turning to page 43 in Mathew Slinn's book "Build Your Own Electric Bicycle – published by McGraw Hill." If you think "pocket protectors" are cool, have been called a "propeller head", are an engineer, or need to teach a college course on electric bicycle building, then I recommend buying that book on www.Amazon.com for around $20. It's written by an Englishman and Engineer that I hold in very high regard for the depth of his knowledge on the subject.

A hearty thank you and congratulations are due to Mathew Slinn and McGraw Hill for producing "Build Your Own Electric Bicycle, the best technical reference book on the subject to date. Reading the book was heavy lifting and it was written to apply to e-bike enthusiasts internationally. That book would make a great college textbook on the subject.

Most importantly and thanks to Mathew Slinn's research, you'll find in his book an upward arching graph demonstrating that the incremental increases in speed are costly. Meaning that even e-bike technology cannot escape the "law of diminishing returns" that we heard much about in college. (Slinn, 2010)

Also, when selecting a power keep in mind your states' Motor Vehicle Laws include a section defining pedal assisted vehicles. In California (and most other states) the e-bike needs to stay under 25 mph maximum speed to stay as an unregistered vehicle. I'm happy with my 600 watts of power on my 20 year old bike frame. The "sweet spot" for "local errand" style bikes is 600 to

750 watts. Check your states Motor Vehicles Laws by going to www.ebikebook.com for links to help research this for US and Canadian users.

I selected the following for my first electric bike, it worked so well that it inspired me to write this book and become an "electric bicycle advocate". Here is the configuration of my first "local errand" e-bike.

- Mountain Bike Frame with 18 gears
- 600 watt, brushless hub motor, in front wheel
  - Brushless hub motors are known for a little less torque than the hub motors "with brushes" but the good news is that there is no annual maintenance issue of replacing the brushes in brushless kits.
- 36 volt battery system (SLA Battery)
- Thumb throttle with power level indicator

Photos of this e-bike follow in about five pages.

I think a 680 to 750 watt power system and a Lithium battery is even a better fit for me based on my weight of 200 pounds. Though, the cost would be several hundred dollars more. To do it over, I would spend the extra $100 for the power to get 680 watt brushless motor, but for this first conversion for me I am very satisfied with the results.

I still shy away from the $500+ Lithium Ion batteries. The price for lithium batteries needs to come down, or maybe I will learn how to source and build a lithium battery pack for my bike. The extra $300 between Lithium and SLA batteries is a good investment in maximizing the riders experience while riding. It's especially worth considering if you want to save 15 pounds off the bike, and get a battery that is much faster charging, provides extended range, and lasts longer in years. Though I am currently exploring options from a lithium battery specialist claiming to save me a couple hundred at www.falconev.com I will publish the results on this source as soon as possible.

My experience with Lithium Ion batteries is limited to the use of cordless power tools... I have owned several Makita cordless drills over the years. I picked up another one last year for our major house remodeling project, which included adding a second story. I purchased a Makita 18V Impact Driver and Cordless Drill with Lithium Ion batteries. It is the most amazing tool I have owned. It felt like 5-10 times the endurance, and torque of any other Makita I ever used or owned. The light weight of the battery is as impressive as it is mind-boggling. The contractors agreed. In fact, a couple of our contractors bought Makita Lithium Ion Tools very soon after they tried mine. So buy Lithium Ion batteries if you can afford it AND be able to use it without worrying about the risk that your $500 battery might get stolen.

One advantage of an SLA battery is security. I like the fact that the bike weighs more with the SLA battery. For example, when I park it outside a sandwich shop and there is nothing to lock it to. I just use a cable lock and lock the rear wheel to the frame. With the SLA battery in the rear, it is not really possible to pick it up and run down the street with it. I don't think a lone thief could quickly lift it into the back of a truck while I'm watching through the shop's window. Though, it might even be entertaining to see someone try. My back hurts just thinking about it!

In summary of Part One – Defining Usage Needs, I feel we are off and running (or biking as our case may be) in the right direction. We have established that the Needs Survey is the first step and critical to satisfactory results on your e-bike project. And the Needs Analysis that we will perform based on the Needs Survey will determine much about what direction you should go in terms of electric bicycles. First, here are some black and white photos of my first e-bike to ponder. Or for high resolution COLOR PHOTOS just go to www.ebikebook.com for the following photos in color.

Following the pictures we will jump into Part Two – Choosing a Kit. Part Two will walk through the Needs Survey questions one by one, in great detail to help you analyze your Needs Survey answers that you should have already completed at least in part. Then you may recall that Part Three is Installation, and Part Four is Time to Enjoy!!

My "local errand" e-bike

Close up of rear rack, folded basket, battery bag and under seat tool bag. (note basket starts at rear axle)

Thumb throttle with battery power indicator

Locate bell for quick use by thumb without looking

Front wiring brought up fork then across to underside of top frame tube. (Note check for enough slack in wiring as to not affect steering)

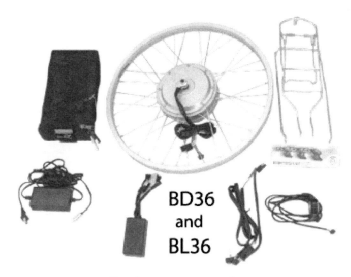

BD36
and
BL36

BD36 - Brushed, Power & Performance
BL36 - Brushless... Durability and Long Range
Wilderness Kit Options

Rear
Mount
Kickstand
(Highly
Encouraged)

Most importantly, this is supposed to be a fun experience! So, don't forget to "crank the tunes" and have a good German beer while building your e-bike! And don't sweat the small stuff!!

Part Two

Choosing a Kit
(And other bike parts)

ANALYSING YOUR USAGE NEEDS SURVEY

GOALS:

What do you want to accomplish?

What you want to accomplish is a very important question and a great place to start. Start your answer with just a free-form rambling of all the ideas and expectations you currently have about e-bikes. Do you want to save money? Do you want to save time? Do you want the greenest and most cost effective motorized transportation solution? Are your knees getting old and you like going places on your bike fast?

Ultimately, you will need to summarize these needs into a specific e-bike style listed below. In other words, what is the primary usage mode? I believe that there are FIVE distinct styles that listed below relate to different primary uses. It's not to say that one bike can not to it all. It's just that the different "styles" have an optimal configuration that is unique from the other styles. And then there is ONE "Combo" style, so SIX styles in total. The six questions to ask yourself are:

1. Do you want it for *commuting* to work?
2. Do you need it for *local errands*?
3. Do you need it for *touring vacations?*
4. Do you primarily want it for recreation (not for errands or commuting to work or touring?
5. Do you want to pack the bike for travel on train, plane, RV, boat or automobile for use at your end destination? Do you need to do "Mixed-Mode Commuting" and pack your bike on a commuter train or bus?
6. Or will you be a heavy user with one bike incorporating *multiple styles combined?*

For the sake of simplicity let's refer to these "e-Bike Usage Styles" as "e-Bike Styles" (or just "Styles") and *the six "Styles" are*:

1. "Commuter" Style
2. "Local Errand" Style
3. "Touring" Style
4. "Recreational" Style
5. "Folding Bike" Style
6. "Combo" Style

Ok, so let's review these *six* e-bike designs now.

"Commuter" Style

There are over two million Americans riding their bikes to work today. (Tuff, 2009) This is a large and very important segment of the e-bike community. I wish I had the opportunity to ride to my primary work. The good news for me is that we have a secondary and family business about a half a mile away. And I ride instead of drive there as much as possible. I walk sometimes too, but usually I'm pressed for time and need to carry some items or tools. So, walking is much less feasible for me, even though the family business is only an eight to ten minute walk. I go there usually on the weekends, so the roads are safer too.

Also, I feel safer on the bike than walking, since I can out run a stray dog, or group of young LA punks. Feeling safer and better prepared to fend off "an attack" is important to me. Welcome to the big city! When out and about I'm sporting a helmet, gloves, and yes years of Karate, Jiu-Jitsu, boxing, and even the slow moving marital art of Tai Chi that I still practice every week at a class. So you caught me, I don't like getting mugged. It's happened to me several times before, and I tend to resist and fight rather than give into a punk's demands. The smarter person avoids the issue entirely, or just gives the punk what they want. It's proven many times easy for me to say, hard for me to do. Hopefully, I have matured on this for next time.

In the Summer of 2010, a friend of mine ("Paul the CPA") and I while on a 25 mile road bike ride in Long Beach stopped briefly to stretch at the turn around point for just a few minutes, as the first half of the ride was into a strong and consistent 15-20 mph headwind coming off the ocean. Within 3 minutes at a busy beach on a highly traveled bike path in the summer, we were targeted for an armed mugging and road bike theft. So surreal the event was, as 5 days before we were at the exact same spot with 700 other cyclists starting the 30 mile Tour of Long Beach ride. Also, we were alongside 100's of private yachts, when the 12 and 15 year old punks approached. We had watched their movements and knew what was up, we got out of there literally with seconds to spare, as they shouted insults to draw us in. The oldest punk was shirtless and wearing a backpack. He had his hand behind his back and holding onto something sticking out of the bottom of the backpack that the younger one just took 5 seconds to arrange and place in the older one's hand. My guess was it was a large kitchen knife. Am I a native Angelino or what? Ten minutes later, I mentioned this to a Long Beach Police Officer at a different Long Beach Marina. The LBPD Officer didn't even blink, said that was no surprise and he would let the on duty officers in the area know. He said that was relatively commonplace in that Downtown Long Beach Marina area. Ouch! Well, we haven't cycled downtown Long Beach again. The take away here is to keep your eyes always open, even when parking your bike. Had we not had our eyes open, we would surely have lost our bikes, phones and wallets.

Bike theft is an important issue (especially for e-bike commuters) and worth repeating, especially after you put several hundred dollars and many extra hours into your new e-bike. I will cover this more in the Environments section of this Needs Survey. According to the numbers from the FBI over 1 million bikes were stolen in the U.S. in 2006! (FBI, 2007) And a recent LA Times article cited a 29% increase in Los Angeles County bicycle theft from 2008 to 2009, and in Downtown LA a 57% increase. (Winton, 2010) So the main point I hopefully drove home here is to commute if you can, but only if you have a safe place to secure your newly built e-bike. Ideally, you'll have an indoor place like your company's warehouse or even in your office, to lock and secure your commuting transportation.

Back to the main topic of this section, the "Commuter Style" is a great way to go to work in my biased opinion. Of all the styles, it's the one that implies the most miles annually compared to the other styles of e-Bike design. It is a very demanding style in terms of power, and battery endurance. My thoughts here are that the average commuter will want to arrive at work without breaking a full sweat. If this is your primary purpose for the bike, you may want to seriously explore your options for buying an integrated e-bike from iZip, Ultra Motor, or Trek. Pricing again is around $2,500. It's much more elegant, but that is a lot of cash. But if you are doing the miles to work, AND have a secure place to park it, then it is definitely worth considering buying vs. building. If you log 3,000 commuter miles annually, you just made $1,500 annually!!

A "Commuter Style" design for an e-bike should have many if not all of these attributes:
- Built on a Hybrid Bicycle frame or Touring Frame. Maybe a mountain bike frame, but that depends on your commute road conditions, distance, and riding preferences.
- You'll want a fast bike to cover the miles, so tall 700cm rims ideally.
- Fenders to protect your clothes should you splash through some water or a pot hole filled with water.
- Ideally, front suspension and a seat shock. So far sounds like a $250-$300 Trek Hybrid that I got my wife several years ago.
- Ideally, all that is needed is just a deep front basket. And a web bungee cord for use on the top of the basket. I like the front basket for commuters, as opposed to rear baskets for several reasons. It's important to have better weight over your front wheels for traction, and handling. Plus if that is your work bag or brief case you will want to keep that where you can see

it, to avoid showing up to work only to find you lost it on the road.

- If more than 3-4 miles to your work, then you should go with a lithium battery.
- If not parking indoors at your work, then you would definitely want that $400 to $500 lithium battery to be easily removable. Since your bike could be parked in the same spot 40 hours a week, every week. That gives the bike thief plenty of time to plan the perfect crime of stealing your bike.
- You'll want features like a bell, to warn kids, pedestrians, and other cyclists as you approach at speed to keep it safer.
- Lights are encouraged, since riding in the early morning or at night is possible.
- An 100+ decibel horn to warn distracted drivers not to pull out in front of you as with all that weight your maneuverability and braking are severely compromised.

"Local Errand Style"

A "Local Errand Style" design for an e-bike should have many if not all of these attributes:
- Built on a Mountain Bicycle frame ideally, or maybe a Hybrid bike frame, but that depends on

your road conditions, distance, and riding preferences.

- You'll want to match your tires to your road conditions. Keep in mind that e-bikes will have a higher speed, and heavier weight.
- Fenders are optional. And downright not required if you live in an arid or semi arid region like Southern California.
- Ideally, front suspension and a seat shock. (So far sounds like a $200 to $300 bike if you need to go out and buy a new one. Or under $100 if shopping for a used bicycle, then www.craigslist.com is an option for you.)
- Since one of the primary local errands is hauling food and goods from the market, a deep front basket, and two folding rear baskets off the back rack. Two "web bungee cords" for the front and rear baskets. I like to carry a third web bungee, just in case I stop at a garage sale and need to pile a box on top of the rear rack or front basket. A couple of extra "old school" bungee cords are always in my bike bag too.
- If it is more than 3-4 miles to your market or there are significant hills, then you should go with a lithium battery.
- Since you will be parking outside use a kryptonite channel lock, and a cable lock to secure the frame and both wheels. Personally I like my 20 pound, cheap SLA battery since I

leave it on the bike while at the market relatively worry free. I secure it with three beefy, industrial zip ties to the rack and it cannot be easily removed. The bag that the three 12v batteries sit in camouflages the batteries nicely and is also connected with short Velcro straps. The zip ties also hold the batteries tight to the rack when I hit bumps.

- You'll want features like a bell, to warn kids, pedestrians, and other cyclists as you approach at speed to keep it safer.
- Lights are encouraged, since riding at night is possible.
- A 100+ decibel horn to warn distracted drivers not to pull out in front of you as with all that weight of your groceries and goods. Remember, your maneuverability and braking is severely compromised by the weight.

"Touring Style"

A "Touring Style" design for an e-bike should have many if not all of these attributes:

- Built on a Touring Bicycle frame ideally. Maybe a Hybrid bike frame, but that depends on your typical touring road conditions, distance, and riding preferences.

- You'll want to match your tires to your road conditions, higher speed, and heavier weight. Kevlar line tires with puncture resistant tubes.
- Fenders are optional, but probably will make good sense for most. And quite frankly required if you live in a region that is known for rain during touring season, like the Pacific Northwest.
- A very well padded and "springy seat" to help with the long miles. (So far that sounds like a $500 to $1000+ touring bike if you need to go out and buy a new one. Or under 1/3 the price, if you get lucky and can source a used touring bike while shopping www.craigslist.com . Or perhaps at half the price on www.ebay.com .)
- Since one of the primary tasks is hauling heavy loads for long distances, you'll want to source the best quality racks, touring bags and baskets you can afford and matched to the type of tours you are planning. I believe my favorite "web bungee cords" for the top of the racks/baskets, would also work amazingly for hauling touring stuff.
- If you want to cover more than 5-10 miles a day with you full load or there are significant hills, then you should go with a lithium battery. Definitely look into the more expensive 48 volt battery systems if your daily range is 25 miles or more per charge.

- Since you will be parking outside sometimes, use two locks often. A Kryptonite channel lock and a cable lock that way frame and both wheels are secure, and the frame is locked in a way that is bolt cutter proof.
- You'll want features like a bell, to warn kids, pedestrians, and other cyclists as you approach at speed to keep it safer.
- Lights are encouraged, since riding in early morning or at night is possible.
- An 100+ decibel horn to warn distracted drivers not to pull out in front of you as with all that touring gear and e-bike weight your maneuverability and braking are severely compromised.

"Recreational Style"

A "Recreational Style" design for an e-bike should have many if not all of these attributes:
- Built on a Beach Cruiser Bicycle frame ideally. Maybe a Hybrid bike frame, but that depends on your typical road conditions, distance, and riding preferences.
- You'll want to match your tires to your road conditions, higher e-bike speed, and heavier weight. Kevlar line tires and puncture resistant tubes are recommended.

- Fenders are optional, but probably will make good sense for most, especially if you live in region that is known for rain during biking season, like the Pacific Northwest.
- Ideally, a very well padded and "springy seat". So far sounds like a $100 to $200 bike if you need to go out and buy a new one. You could always try your luck at sourcing a cruiser from www.craigslist.com . What works for me is to check in the evening after dinner for brand new posts in the immediate area, and have your cash and bike rack/truck ready to go on a minute notice should you strike a deal over the phone with the seller. Also, Target Stores have a great value on bikes in this category, as do Costco, Amazon.com, and Toys R' Us.
- Since one of the primary goals is having fun or recreation with this type of e-bike. Get a front basket for your towel, volleyball, book, lunch, iPod, iPad, iPhone or whatever else you want to carry in a backpack, and of course don't forget my favorite "web bungee cord" to secure it all.
- If you want to cover 20 miles a day then you should go with a lithium battery. A SLA battery is probably good for 10 miles of pedal assisted riding on a moderately flat trail.
- Since you will be parking outside sometimes, use two locks often. A Kryptonite channel lock, and a

cable lock that way frame and both wheels are secure.

- You'll want features like a bell, to warn kids, pedestrians, and other cyclists as you approach at speed to keep it safer.

"Folding Bike Style"

Have I taken this "e-bike Styles" discussion too far? I say confidently, "aux contraire" and that folding bikes will be a significant percentage of the e-bike market in the future. Today, two million Americans ride their bikes to work. From 2007 to 2008, the number of New Yorkers that rode their bikes to work increased by 35 percent! The term "mixed-mode commuting" has been coined. In New York while riding on the Greenway practically every third bike is a folding bike. (Tuff, 2009)

The message here is that Folding Bikes are a hot and growing segment. Personally, I love my Dahon folding bike, it easily fits in my Suburban for family trips. I use my Dahon foldy for recreation, and local errands while on a family vacation. It is not an e-bike, and I have no plans to convert it. It's lightweight, and transportable. I did attach a basket to the rear rack without compromising the folding. It can for the most part handle hauling a couple bags of groceries stuffed into a duffle bag and bungeed to the rack from a store 5 miles away. I have time to bike slower on vacation and bike for recreation, too. For me it's a great 10-12 mile ride on a foldy while on vacation. Whereas in perspective on my road bike, I like 20-30 mile rides .

So, yesterday I saw a folding e-bike cruise down the sidewalk in our small main shopping district, defintely not as fast or sturdy as my mt. bike, but it was much lighter and was moving along nicely. I did not see any rear or forward baskets. But the battery did look removable, and the bike looked like an integrated e-bike, not a kit. I bet it could carry only half the groceries or less of what my mountain bike frame e-bike could carry. Folding bikes usually have 20 inch wheels and that wheel size is not ideal for the speed you can generate with a 600 watt plus e-bike motor. You'll feel safer with that kind of power on a mountain bike or hybrid bike. Folding bikes can make good e-bikes. A Lithium Ion battery would keep the weight down, and something around the 400 watt range for the hub motor would make a nice entry level folding e-bike weight and speed combination. If you are an experienced rider, then you should be able to handle up to 600 watts on an folding bike. But remember those frames are made to be light weight, they are not designed or optimized for carrying three large brown paper bags of groceries several miles every week. Folding bike frames will not be nearly as safe at high speed as a mountain bike, especially when you hit a significant bump on the road.

A "Folding Bike Style" design for an e-bike should have many if not all of these attributes:

- Built on a Folding Bicycle frame. Wheel sizes vary depending on your typical riding needs, road conditions, distance, and riding preferences, but I think e-bike conversion *kits* only go down to 20" wheels.
- You'll want to match your tires to your road conditions, higher e-bike speed, and heavier weight. Kevlar line tires and puncture resistant tubes would definitely be in any e-bike style.
- Fenders are optional, but probably will make good sense for most. I have fenders on my Dahon folding bike, they work great.
- Ideally, a very well padded and "springy seat". So far sounds like a $250 Dahon folding bike if you need to go out and buy a new one. Check out www.sunrisecyclery.com that is where I got my Dahon. I think through www.Amazon.com . You could always try your luck at sourcing a "foldy" from www.craigslist.com . What works for me is to check in the evening after dinner for brand new posts in the immediate area, and have your cash and bike rack/truck ready to go on a minute notice should you strike a deal over the phone with the seller. Of course you might get lucky on www.ebay.com but that was a pretty weak source for folding bikes a couple years ago when I was searching.

- Since folding is important, you should be able to attach at least one folding rear basket to the rear rack without compromising the folding.
- If you want to cover 20 miles a day then you should go with a Lithium Battery. 10 miles is a good number for the old cheap SLA battery.
- Since you will be parking outside sometimes, use two locks often. A Kryptonite channel lock, and a cable lock that way frame and both wheels are secure.
- You'll want features like a bell, to warn kids, pedestrians, and other cyclists as you approach at speed to keep it safer.
- Lights are encouraged, since riding in the early morning or at night is possible.
- An 100+ decibel horn to warn distracted drivers not to pull out in front of you as with all that weight your maneuverability and braking are severely compromised.

"Combo" Style

Choosing a Combo style may seem ideal, but it's important to note that Combo's are not optimized for a particular style, but it of course combines multiple styles. Hopefully, you can do this and not compromise your primary usage type needs. For example, a Folding

Bike that is used for local errands will not be nearly as safe under weight, nor will you be able to haul as much.

For your first e-bike conversion you may want to consider lowering your expectations about how you plan to use, and focus on just *one* of the first five styles. You can optimize the bike for one style and get the best configuration for your needs. In other words, don't try to do it all only marginally well, when you could create a design to do a more limited scope really well.

I decided to build my e-bike following a "Local Errand" style. I do not use the e-bike for commuting to work or school. It is not a folding bike, and I do not use it for recreational trail riding. I have a road bike and a hybrid bike for organized bike events, fun tours, or cruising at the beach trails, which I like to choose from depending on the location distance and who's going.

Again, you should decide as a result of the Needs Usage Survey on the "e-bike style" that best meets your needs of how you will *primarily* use it. Consider narrowing the scope and focus of what you want to accomplish and your design will benefit greatly.

What I find surprising is how much fun my e-bike is to use for local errands. I love quietly zipping around cars (parked or moving) in the parking lot, leaning into a fast turn, or cruising down sidewalks to pass cars stuck at traffic lights/intersections. I can park the e-bike usually right in front of the store, usually closer than any car. Basically, driving a car and waiting at traffic lights is boring. In other words, why drive for a local errand, when you can ride an e-bike!! This is why I chose the "local errand style". I feel that whatever style you chose that you will find it fun. Given that most people who will read this book work hard at life and careers, I would like to encourage again choosing the e-bike style that you will use the most and *enjoy* the most.

What is driving you interest in e-bike technology?

Is there something driving your interest that has a timeline or target date associated with it? What is the timeline that you are giving yourself to finish the project? What is a primary goal that you want to accomplish, and have a strong desire to accomplish?

Here's the way I answered the question as it pertained to me. I want to replace 50% or more of my local errands with my e-bike in an effort to save money, lose weight, save miles on my cars, enjoy the bike ride, work on my bike rider tan, feel the wind in my face, and set a green transportation example for family, friends and neighbors.

This will help you with selecting your design style, and being satisfied with the results. That is you don't want to forget to address one of your key desires that is driving you towards e-bike technology. Since there are literally hundreds of decision and factors to consider, let's make sure we don't lose sight of the main ones.

Do you want to save money by reducing the miles driven for local errands?

If this is one of your primary reasons for building an e-bike, then that helps you finalize your design on a "local errand" style. Your decision on motor power, battery, rack, baskets, etc. should include what you need to accomplish "local errands" in your environment.

This means that you will need a heavier frame, baskets, more watts, and more battery juice. Keep in mind that the extra weight of groceries, especially if hills are present, will drain the battery much faster and shorten your range.

For me stores are 2-3 miles away using side streets and I have a significant number of hills. The configuration I put together has worked great for local errands in this environment. I have three baskets (one in the front and two folding baskets attached to the back rack). The configuration is a 600 watt front hub motor and a 36volt SLA battery. All this built on a mountain bike frame using the local errand design style. I bet I would like even better a 680 watt or 750 watt front hub motor, and a 36v lithium battery, but those upgrades would have totaled several hundred additional dollars.

I feel that I optimized my value at the 600 watt motor and heavy SLA battery by saving those several hundreds of dollars and purchasing the kit for under $400. I was able to save money by doing the installation myself, as well. This reduced the amount of time it would take for my new e-bike to pay for itself with savings generated by not driving my car for local errands.

Do you want to lose weight, burn off stress, and lower your cholesterol?

This also was one of my goals. For me, the 45 year old software sales rep with three teenagers, these are very important goals. I lost 20 pounds since the start of the year and have 20 more to go till I hit my ideal weight. The Mayo Clinic points out that excess weight helps bring about high cholesterol and of most concern is the increase in the LDL (the bad cholesterol). But losing five to 10 pounds can significantly lower levels of your LDL. The Mayo Clinic recommends a healthy diet along with regular exercises. It recommends getting a doctor's approval first. Ideally on most days of the week you should exercise 30 to 60 minutes a day, for example a brisk daily walk, bike riding, or swimming laps.

You don't have to do the 30 to 60 minutes of exercise in one session. You can do six 10-minute intervals of exercise sessions. Increasing physical activity, losing excess weight, eating healthy foods and kicking the smoking habit if you smoke are lifestyle changes that will lower your LDL (bad cholesterol) levels, improve your HDL (good cholesterol), and help you avoid cardio vascular diseases like heart attacks and strokes. (Mayo Foundation, 2010)

The message here is that the more you use your e-bike, then the more you improve your chances for significant improvements in your overall health. *Design an e- bike that will maximize the number of times that you will use it* and your benefits will be more than just financial. It could save your life! Just be safe, always wear a helmet, and watch out for the cars!

Do you want to improve your aerobic conditioning?

This is another great reason to add miles of bike riding to your weekly totals than what you are currently riding. In other words, if you are only *replacing* miles you used to do on a regular bike with e-bike miles, that means you will be sweating less for the same number of miles and likely not getting as much aerobic conditioning as previously. Keep a journal of your miles to inspire and motivate yourself. Create a goal for yourself that includes you adding miles to your weekly totals above what you did before you had an e-bike. Track miles in your journal of your regular and your e-bike distances. To do this well, just add a $10 speedometer (aka odometer) to your e-bike.

Do you want to set an example for other "automobile zombies" that it is possible to run errands on a bike without breaking a full sweat?

I always have believed in and practiced leading by example. "Do as I say, but not as I do" is the wrong way to raise a family or participate in your community. The true environmentalist does not drive a car for a local errand. I have become an electric bike advocate. Not just in words or by publishing a book about it, but in actions. In fact speaking about being an advocate, on my website www.ebikebook.com my Mission is "Build *a Community that Promotes Electric Bicycles as a Green Transportation Opportunity.*" If that's not the website of an e-bike advocate, then I don't know what is.

If you answer yes to the above question, then your style is not just "recreational". You are someone that wants to use this new green transportation technology as much as possible, which is for "local errands" and perhaps "commuting", too.

If you bike for local errands and/or for commuting, you are absolutely a "cyclist". As the infamous blogger Bike Snob NYC wrote in his 2010 book, the definition of a cyclist is "one who rides a bicycle, even when he or she doesn't have to." He goes on later in his book to make a very important and very relevant point. Eben Weiss, the BikeSnobNYC, wrote, "I believe that by simply being cyclists and riding our bikes we're actually doing as much for cycling as anybody. This is the beauty of cycling – all it really takes to be a cycling advocate is to ride your bike. The more people who are out riding their bikes, the more cycling benefits." Thank you Eben, I could not agree with you more! (Weiss, 2010)

So, the more people that cycle, the more cycling benefits. If you are not convinced yet, then I offer an example. In 2010, Los Angeles had its first cicLAvia event for 5 hours. There were an AMAZING 100,000+ cyclists and participants that showed up for this inaugural event!! It was also the only event of 2010 of its kind in LA. What a fantastic event, and not one "incident". With 100,000 in attendance, cycling absolutely benefited. In 2011, three such events are scheduled (April, July and October), and plans for 2012 are targeting as much as a monthly cicLAvia event. See www.cicLAvia.org for details.

As a father of three, soccer/volleyball coach for the past dozen years, and scout leader for seven years, I have always had this drive to *lead by example*. I do gain a satisfaction from being an innovator for something like this green technology and I jump at the chance to explain it to any one that shows an interest or asks about it. So, be true to your personality as you will get many questions from strangers asking you about your new innovative bike. Ask yourself...Am I an innovator at heart? Do I like to lead and teach? Will it cause me stress if people ask about my bike? Will this issue affect how I use the electric bicycle and how? Currently, anyone that is caught riding an e-bike today will get multiple questions daily about the e-bike conversion technology, as I have recently found out. In fact, I had so many questions, that I felt the need to write this book and carry my www.ebikebook.com business card.

Do you have a budget for this e-bike project?

This is a very important aspect to selecting the right conversion kit or integrated e-bike for your needs. Everyone has a budget in mind when they enter into this. My budget was "under $400" for the kit and do the labor myself. I then put another $50-$100 into the bike in the form of a front basket, web bungee, speedometer, tire air pump, and new water bottle cage (my 20 year old cage was looking pretty sad). That's a realistic budget if you want to build a "local errand" bike. This assumes of course that you have a bike in good working order prior to the e-bike upgrade.

If you want a "commuter style" e-bike, then you have a lot more miles and cost savings to justify a more expensive solution, and hence a larger budget. Again spending up to $3,000 is not uncommon for a fully integrated commuter e-bike with a 750 watt motor and lithium batteries integrated and hidden into the frame. The "commuting to work" problem is a bigger problem to solve and you should expect to spend more if this is your challenge and you work more than 3-4 miles away from home. At a minimum you should be at the $750 budget, so that you can include the much lighter and longer lasting lithium batteries.

Recently, I came across a name new to me for electric bicycles. They have a unique position in the market place. They are bundling a 48 volt lithium battery into the mix. This makes the bike an incredible fit for the commuter style. They are claiming a 55 mile range without pedaling, and 75 mile with pedaling! The company name is Busetti. That bad news is that a replacement battery is $900!!!

Searching the web you'll likely find websites and stores pricing to be around the $1,200 mark for a very powerful 36v e-bike with full suspension. This is a very tempting price point, but is it a good value? See my website www.ebikebook.com and "fan page" to get the links to the latest distributors that appear to offer the best values. My forum will help keep these vendors honest with the latest customer generated reviews.

RIDER(S):

Are you an environmentalist by nature?

An e-bike at this point is more than just transportation; it is a "green transportation opportunity". It is something that is very innovative and likely soon to be very popular in the US like it is already popular in many other countries and across almost all continents. Again this means, other environmentally minded people will approach you for info on your experience with this new "clean air" technology. They may even consider you an environmentalist just because you ride an e-bike. If you hate talking to an environmentalist, you may want to consider buying a motorcycle or a moped.

The practical implications here are that an environmentalist will want to use the e-bike for more than just recreation. The environmentalist would strive to use the bike for "local errands" and even for "commuting", if possible. In other words, if you are an environmentalist, that will impact how you will want to use your bike. Environmentally thoughtful people will do their best to replace whenever possible automobile miles with electric bicycle miles. To be "greener" think about what trips by car you could replace with your e-bike. Does the design you have in mind fit with that?

How many years have you been cycling?

There are many reasons for asking this. Overall you want to consider your cycling experience and weigh that against your usage goals. The style of bike that you build or buy needs to be compatible with your cycling ability. The weight, speed and environment can greatly impact your safety and use of the bike. For example if you are a *novice* cyclist, then you should opt for a lighter weight battery (lithium if you can afford it), a less powerful motor (500 watts to 250 watts) and stay off the streets as much as possible.

Also if you are a novice rider, that implies you are a novice at bicycle maintenance issues as well. Building an e-bike is less desirable than paying more for an integrated e-bike that is already built and ready to use. Integrated e-bikes often have a pedal pressure activated throttle so it is less for the novice rider to think about. Most kits including the one that I installed have a "thumb throttle" or a "grip throttle". Shifting gears while operating a thumb throttle is tricky, especially if you want to keep the same motor speed. A twist grip throttle is easier to handle while shifting gears. There is value to pedal pressure activated throttles offered by many of the integrated e-bikes. Though, I do like hitting the throttle in the flats while not pedaling to quickly accelerate and cruise along near 20mph. And a pedal assist throttle means you have to always pedal, something I definitely don't do because often I am going faster than 17th gear.

Do you have experience riding a bike that is loaded with extra weight? (i.e. pulled a bike trailer or "tag along", handled a newspaper route as kid, loaded baskets/rack, driven a friend that sat on the handle bars or rear rack, etc...)

As unrelated and inconsequential as all those cycling experiences that appear above are, they all do relate to handling a bicycle with more than normal weight. They are all things that I have done for many, many years for thousands of miles. So for me handling the significant additional weight was easy. The toughest part of the extra weight for me is maneuvering the e-bike when I need to stand it up on the back wheel to get it up one step and out the back door of my garage.

What is relevant is that riding a bicycle loaded with an extra 40 pounds of motor/battery, and extra 30-40 pounds for rack, baskets, and market goods is an additional 70-80 pounds. If you are building a local errand bike, then you need to consider your cycling ability level and conditioning. If this is a concern for you, then a lighter configuration of the e-bike with a lithium battery is the way to go. If in a 36 volt battery configuration you could save a significant 18 pounds off the bike just for going to a lithium battery as opposed to a SLA battery.

Have you owned a motorcycle or moped? (or are proficient at driving a motorized two wheel vehicle)

I think of my e-bike as a very light, sidewalk legal, quiet, clean air, and slow...motorcycle!! It's safer being on the sidewalks and side streets for me than wrestling with other automobile drivers at 40 to 50mph on city streets racing to the next red light. If you feel comfortable with the steering, weight and braking of a motorcycle, then you will absolutely feel comfortable with an e-bike. An e-bike is really most similar to a moped than to a motorcycle. The point here is that you can handle the extra weight of an SLA battery, if you feel very comfortable behind the handle bars of a motorcycle or moped.

How much do you weigh?

Your weight factors into several things. I weigh about 200 pounds. My mountain bike handles that fine. Add 45 pounds of battery, baskets, controller and motor. Then I add about 30 pounds of groceries. That's 275 pounds on a bike frame. I don't think the frame would be happy with much more. With the baskets loaded I avoid hitting big dips in the road at 20 mph. The fear is not just losing some grocery items, but more importantly to prevent cracking the bike frame on my 20 year old Diamondback mountain bike. The bike handles the weight well, but "caution is the better part of valor".

Your weight is a significant factor in the mileage range of your e-bike and taking off from a stop. There is a huge difference in results if you weigh 100 pounds versus 200 pounds. If you weigh only 100 pounds you can put together a lighter weight frame and you may want to opt for the lighter lithium battery. It sounds counter intuitive let me explain. At 200 pounds the 24 pound battery is only 12% of my weight. When I ride I barely know it is there until I stop. For the 100 pound rider a 24 pound SLA battery is 24% of their body weight. This affects the feel and weight of the ride much more.

Lastly, I recommend that everyone keep a journal of their miles on the e-bike and their weight. This will be motivational for rider and others. Ultimately on www.eBikeBook.com I would like e-Bike users to publish their results in terms of miles they rode. The only benchmark I know so far is from one person that lost 20 pounds since he started commuting to work on an e-bike. He logged about 3,000 miles in that one year he lost 20 pounds.

Do you have lots of tools and consider yourself handy with them?

For me the answer was easy. I've always been a "do-it- yourselfer". This made it easy for me to decide that adding a conversion kit was the most cost effective way to go for my first e-Bike. This almost goes without saying, but if you do not have tools and are not handy then the best solution is an integrated e-bike. Although it's not rocket science, you could have a friend experienced in bike maintenance help you out.

Do you do some basic bike maintenance yourself?

This is an important question. Think back about bicycle maintenance issues, especially little things like adjusting the brakes so they hit properly on the wheel and at an ideal distance on the hand lever away from the handle bars. Do you do that type of maintenance yourself and do it well? Or do you leave that for the bicycle shop? The point of this question is that you will most likely need to readjust your brakes after installing the new motorized wheel. I was recently surprised when chatting with several bike techs at one of West LA's oldest and largest bike shops that none of them have every worked on an e-bike before. So if you are installing a conversion kit versus purchasing an integrated e-bike. You will likely be best served by doing most of the maintenance yourself.

Will you be the only rider?

This seems like an obvious question and easy answer, but dig deep here. The issue is that the motor power, battery power, bike weight configurations of e-bike conversion kits can vary greatly. A configuration may not meet all your needs and not give you the power or distance that you want. You could wind up losing several hundred dollars and time.

For example, let's say you have a 1 year old and you like to ride for recreation with friends and family. Well your 1 year old is not going to ride the e-bike, of course, but you may log 1,000's of miles with a kiddie trailer in tow. My three kids took their turns for a few years in our beloved Burley trailer, singing, talking, or sleeping as we trekked to one of their favorite beach spots or parks. Often two kids at a time and by hooking up a "tag-a-long" in between the trailer and your bike, you could take three!! That actually works incredibly well, one kid would help pedal, so it was like hooking a trailer to a virtual tandem bike. Needless to say I would have a coffee with breakfast on the weekends as that was a lot of work with 3 or 4 of us on one long bike.

The main point of this example is to consider how your usage needs impact your conversion kit decision. Weight and hills drain batteries very fast, so does starting from a dead stop without pedaling. Pulling a trailer is definitely weight. If the manufacturer of conversion kit is stating up to a 20 miles range. Cut that by a factor of THREE or FOUR. Meaning that speaking from my limited experience, I would advise you to expect a 36 volt SLA battery to help you near full power for 4-5 miles when pulling a trailer and peddling. Though, maybe you could get 7-8 miles if you went half throttle, peddled harder, and didn't start and stop often. Plus you could add 2 miles of low power use before the battery totally stopped assisting. When the battery is low my Wilderness Kit drops to a low power mode which uses fewer amps per hour from the battery and extends the range. Of course this means you have half the power from the motor.

I would suggest at minimum a 600 watt motor if you are pulling a trailer and opt for the more expensive lithium battery. The lighter battery will save the 20 pounds to make the bike safer; making it is less likely to fall with tragic results on your toddler. The lithium battery will charge faster, have longer range, and offer more years of use.

Here's an interesting issue that I thought about today as I was putting away my e-bike in my garage. I'm an athletic guy of at least average strength for my age. I like to hit the gym and free weights 2-3 times a week. With the extra 45 pounds of motor, SLA battery, converter, rack, and three baskets it's a little challenge to walk my "local errand" style bike around on the rear wheel only while holding the handlebars. I don't think my tall and strong 120 pound wife or my two younger kids today would be able to put my e-bike in the garage the way I do because of the step, and the weight of the bike. Going up or down that step like most steps requires balancing a heavy e-bike on its rear wheel. So, again the lithium battery offers a lot of value. Going forward, I think all my e-bikes will be lithium batteries.

If not the only rider, then what is different about the other rider(s) in terms of skills, physicality, and needs?

If the other riders are much less skilled at cycling then you'll want to keep the bike as light as possible. For less skilled riders, I would encourage starting with a "recreational" or "folding bike" style as those are much lighter than the typical "commuter" or "local errand" style. Also, with a "local errand" style implies the rider will need the strength and skill to handle the bike loaded with two or three baskets filled with groceries.

For the novice cyclist riding a bicycle with an extra 80 pounds on it does require more skill. If you are a novice rider strap 20 pounds to your rear rack and get the feel of it on your regular bike. Now picture or try doubling that weight to account for 15-20 pounds of groceries. Keep your safety and the safety of others in mind when deciding on the style of bike you want to build. Novice users should make the bike as light as possible, and opt for lithium battery power. Also take into consideration how the other riders plan to use the bike for recreation, commuting and/or local errands.

Have you ever felt like MacGyver from the TV show (or MacGruber from SNL)? Do you want to be in a position where you might need to re-engineer some components or settings on your bike?

All kidding aside, this is an important question. Re-engineering a pedal bicycle to motorized vehicle is fraught with unseen challenges. For example, I had to use a Dremel, which is 9,000 rpm hand tool with a diamond dust bit, to grind a wider opening in a couple of pieces to fit the new, thicker axle bolt. The pieces I modified were just safety washers with an extra hole that allowed a braze on bolt to secure the front wheel axle should I lose the bolt on the front wheel. I could have left off that safety washer, but I liked the idea of doing all that I could to prevent a quick trip over the handle bars should the front wheel fall off. Also, I needed to have my son help me mount the new motorized wheel to the front forks. It was definitely a two person job as we had to pull the forks apart a little wider than they had been before. The rim was wider than previous so the front brakes required several adjustments to work well again. The point here is that if you are not mechanically inclined, or have some friends or family "on call" that are able to help if needed, then you should buy an integrated e-bike.

BIKES:

What other bikes do you have?

This will also help you pick your style. When I started out, I purchased a few weeks earlier a used $300 Motiv aluminum frame mountain bike that was 2-3 years old for only $65 on Craigslist with the full intent of using that bike for the conversion kit. Unfortunately, the motor on the front wheel hub and axle was too wide to fit on the thicker aluminum and suspension forks. Luckily, I had a second mountain bike hanging from the rafters of my garage that the kit was able to fit onto. It was a 20 year old Diamondback mountain bike that was already rigged with a rear rack and rear baskets for local errands. I used it for local errands, when time and sweat weren't an issue.

If you only have one bike today, then it is probably time to get a second bike. Try Craigslist.com for a quality bike for cheap. Lots of great sources for quality bikes for cheap. Better yet, check out your local Costco and buy a brand new hybrid or mountain bike for the

$200 range. This way you still have a real bike. The reason being that my electric bike is designed to do one thing for me, reduce the number of miles I drive for local errands. I like to cycle for recreation as well. In fact yesterday in LA on 10/10/10 was the first CicLAvia in Los Angeles, and my family joined me and 100,000 other Angelinos as we rode a 7.5 stretch of roads closed to cars for 5 hours (that's a 15 mile round trip). Check out www.ciclavia.org for more details on this event. There were some e-bikes there, but I would guess less than 500 of the 100,000. I left my e-bike at home, so I could ride slowly with my family on their regular bikes. Heck it would take 2 or 3 of us just to get my e-bike in and out of my van.

Most cyclists that have an e-bike will have more than one bike. In other words, I do not see many cyclists that will have only one bike, and that is an e-bike, especially if it is an e-bike built from a conversion kit. I could see some cyclists having just an e-bike if it is a lighter weight, pedal assisted e-bike from Trek or iZip. As my father always said "diversity is the spice of life". So I say to my fellow e-bike enthusiasts, I hope you have a big garage! Meaning that I encourage you to have and use more than one bike, so long as you have the space for it. If not, then perhaps an e-bike from iZip or Trek is the best way for you to go.

Do you have a mountain bike or hybrid that is not getting much recreational use?

If you do then you are already a couple steps ahead of the game! And your budget can be smaller. This seldom used bike could be the starting point for your design.

Do you currently have a bike with a rear rack, and/or baskets on it?

If so, then this would be a great place to start. Again, your budget to start can be smaller. Also, this will help you build my favorite style of e-bike, the "local errand" style.

Do you currently have a bike that you use for local errands?

If you do, then how far are your typical errands? Do you need to keep you local errand bike a non-motorized bike and why? I was hesitant to convert my favorite "local errand" bike to an e-bike. Now I would definitely do it again, as it worked out really well for me.

If yes, then how is that bike working today for local errands? Does the bike have the ability to manage the extra weight (about 40 pounds for battery and hub motor). Do you hear "creaking noises" from frame when you push down hard on a pedal?

My Diamondback mountain bike has always felt like a stiff frame and still does for the most part. But once I added the 40 pounds and then load up 3-4 bags of groceries etc, I do feel that the frame is about maxed out. I even slow for bumps when my baskets are loaded, that I would normally hit at 20+ miles speed on a downhill before the upgrade. I do that to protect the integrity of the frame and prevent a major failure and crack in the frame as well as prevent damage to my cargo.

When I ask about creaking noises, I am referring to the noises that come from the frame itself. I have often heard creaking noise on weaker frames as they come under load or weight. Frames that make noises are less strong, less stiff, and more subject to a major failure (falling apart over a bump at high speed). If your bike frame makes creaking noises, you should consider a different or new bike for your e-bike conversion.

Do you have a bike with suspension forks, shock absorbing seat, and/or rear suspension?

Suspension makes a lot of sense for any e-bike as you will be traveling at higher speeds than just a normal bike. Without suspension you will feel even the smallest bumps in the road. If your bike does not have suspension forks or a suspension for the rear wheel, then a very cost effective way to add suspension to your bike is by adding a seat post that has a tension adjustable spring for shock absorption in it. I just bought a used one for $15. Try ebay or craigslist. Be sure to measure the size of your seat post with a micrometer or better yet the size in millimeters is usually stamped on seat post tube. Better yet just take the old seat post to your local bike shop to purchase the new suspension seat post. Or maybe even just a seat with springs is all you want. A springy seat is all that I needed to help on this issue.

How many gears do you have on the bike you are targeting for the upgrade to an e-bike?

You might be a little surprised about the gears that need to be on an e-bike. Here are my experiences regarding gearing needs and use of an e-bike. The first thing I realized is that changing gears with my right thumb and holding down my spring loaded thumb throttle at the same time is impossible. A grip throttle would be less of an issue in this regard. If a grip throttle fits on your handle bars then it is recommended over a thumb throttle.

One of the most interesting revelations when discussing what it is like to own and use an e-bike relates to gearing. I leave the bike about 98% of the time in the 17th gear out of 18 gears. I don't shift gears. I just go everywhere fast. Instead of shifting gears, I just give the e-bike motor more "juice" or not. On the steepest of hills of which there are many in my neighborhood, occasionally I just stand up and pedal while giving full throttle till I reach the top of the hill, much faster and much easier. If I do switch gears now it is only with my left hand shifter, which drives the front three sprocket gears. I only switch the left hand driven front sprocket for the very, very steepest of hills and I only switch the front sprocket to the middle or as it is in my case the 11th gear. I don't touch the right hand or rear gears. So I use only two gears, 11th and 17th. How's that for a mind *shift*? (Pun intended)

Of course if I ran out of battery juice, then I would be using all the gears and probably more than on a regular mountain bike because of the weight of the dead battery. Mountain bikes and their lower end gearing is a great fit for e-bike conversion. The lower range of gearing is a perfect fit, not to mention suspension and strong frames. The lower gears are needed to propel the heavier bike should you get stuck without any battery juice. Another reason mountain bikes make a great place to start is that if you need to buy a bike, there are a large number of quality mountain bikes available for quite cheap on www.craigslist.com . Remember the aluminum frame mountain bike I got. Ultimately I did not make it an e-bike, but I only had $65 in to it, and it was not lost as my younger son now has a "new" mountain bike.

How strong is the frame on a scale of 1-10, with 10 being a new $800 mountain bike with front and rear suspension and a 1 being a rusty 20 year old bike?

The frame strength relevancy will vary depending on what e-bike style you want to build. That is quite obvious, but it is of critical importance and needs to be a key factor in your final design decision. If you are just building a "recreational style," the frame strength is not that important. If you want to build a "local errand style" with 2 or 3 large baskets, then the frame strength issue is very important for your personal safety. Also take into account if the frame has suspension, as suspension is preferred.

How old is the bike?

Again this comes back to an issue of safety. If it is an older bike then you might encounter some frozen parts. For example, on my 20 year old Diamondback I could not loosen the rear brake bracket for the hand grip on the handle bars; basically the Phillips head of the bracket screw easily stripped as the screw was "frozen" after 18+ years of never being touched. My Diamondback mountain bike is near its 20th year and in fantastic shape. I always store my bikes in the garage and maintain them with lots of silicon spray literally everywhere. I used to prefer chain oil only, but that creates a bigger mess on the chain and rear wheel. Dirt sticks to oil much better that to silicon. I often use both silicon and oil at different times on the same bike, and much less oil than I used to.

One of the other reasons that my mountain bike lasted so well over the years is that I have several other bikes that I share mileage on. Half the miles I rode in my life were likely on a road bike I had during my college years. I wore that Raleigh road bike out, upgraded the components then wore it out again. The point here is that the bikes age is not as relevant as how the bike was stored, maintained, and like an automobile...the mileage used. Well at least this is what I tell myself when I look at the "old guy in the mirror."

What condition is it in?

Be conservative here. If you overestimate the condition of your bike, then your safety and the safety of others is absolutely at risk. Remember that you are traveling at high speeds with more weight. Faulty brakes or a major failure of your frame could even be fatal.

The following facts are based on analysis of data from the U.S. Department of Transportation's Fatality Analysis Reporting System (National Highway Traffic Safety Administration). A total of 714 bicyclists were killed in crashes with motor vehicles in 2008. Bicyclist deaths were down 29 percent since 1975, but were up 14 percent since 2003. **Ninety-one percent of bicyclists killed in 2008 reportedly weren't wearing helmets.** Thirty-eight percent of bicyclist deaths in 2008 occurred at intersections. Only 35 percent of fatalities occurred on minor roads, meaning that 65% of fatalities occur on major roads. The most dangerous time period (1/8 of a day) is when 21 percent of fatalities occurred in the 3 hours, from 3pm to 6pm. (Insurance Institute for Highway Safety)

To summarize the above facts, ALWAYS wear your helmet. Remember you will be traveling faster. You should always wear cycling gloves as they help prevent injuries when falling. Gloves also help prevent nerve damage in you hands while riding. And whenever possible you should avoid or take extra care at intersections, major roads, and from 3pm to 6pm. As an automobile driver of 12,000 to 15,000 miles annually, I must concur that I tend to be more alert during my morning commute or lunchtime drives, than on my drive home after a long day at work.

How are the brakes, tires, tubes, chain, gears, shifters, seat, grips, and crank shaft?

I of course looked at all these things closely when building and optimizing my new "local errand design". The brake pads were at least 10 years old. It was a great time for new ones. This is a very cheap and yet critical upgrade for your old bike to improve the stopping and safety of the bike ($4 per wheel at www.amazon.com). This came quickly to mind as the old pads made a fierce noise on the new motorize front rim.

I also replaced the tires and added thorn resistant tubes. I used the tires from the new used mountain bike that I purchased on Craigslist. The tires were big and relatively slick and more aerodynamic when compared to my old and typical off-road knobby tires. The new tires were perfect for the dry roads that are 95% of my rides. But in a drizzle or light rain I need to actually let off the throttle in turns.

I cleaned thoroughly any old oil gunk and thoroughly applied silicone to everything and on the moving parts at least twice. I do like seat post shocks, and have several of those. Though, I did not have one that fit the diameter of the frame's seat post tube. But that $65 used Costco bike was a great source for parts again. That bike had an ultra padded seat, and even had springs in the seat with up to an inch of travel. That did the trick, because without suspension on my new e-bike and at those higher speeds my familiar local roads felt bumpier.

What are the measurements of the wheel? (26", 700cm, 24", 20")

For the averaged sized adult you will find most mountain bikes are 26". My folding Dahon bike is a 20" wheel and quite fast. Wheels that are 20" will cost you speed and likely affect "road feel" and stability, especially if you are hauling groceries. A 700cm wheel is the wheel choice for hybrids and commuter bikes. It's a little taller, skinner, and faster than the 26" wheel. Road bikes (used for cycling road races) of course use a 700cm wheel that is a little skinnier than the hybrid/commuter wheel. I do not recommend an e-bike conversion on a lightweight race bike frame for many reasons. Though if I had a Touring road bike frame, which is a road bike engineered to carry weight, I would be tempted to build a fast long distance Touring road bike with a range of 50 miles or longer.

What is the width of the front forks at the axle line?

It is very important to accurately measure this length. I would even measure this on the bike you plan to install the conversion kit on three times. Measure the width of the front forks at the axle line once with the current wheel on and once with the wheel off. Then measure it a second time again while the wheel is off. If not 3 inches in width, see if you can stretch the forks out to 3 inches clearance for a hub motor. Take care to not go beyond 3 inches and avoid breaking your forks when testing if they can expand out to 3 inches. The Wilderness Kit that I purchased requires 3 inches at the front fork axle bolt line.

What is the width of the front forks at 5 inches above the axle line?

This is one of the leading issues that have prevented the successful installation of many conversion kits. Especially susceptible to this issue is the Wilderness Conversion Kit, which is the kit that I purchased. The aluminum frame mountain bike that I bought on www.craigslist.com for $65 to be my e-bike did not meet the criteria of having 3.5 inches of clearance on the forks at 5 inches about the axle line as it had very thick fork tubes and a clearance of only 3.3 inches. That is why I used my older bike.

There are many conversion kits. I wanted the least expensive and most popular I could find as reviews and feedback from users were not easy for me to find, nor plentiful. There are many kits that cost a little more than the Wilderness Kit one that I got, that require only 3 inches clearance at 5 inches above the axle line (not 3.5 inches like the Wilderness Kit does).

This above mentioned situation and issue of poor review information inspired my writing of this as a book to help new e-biker's have a more successful first e-bike design and experience. Part of my strategy is to use my software skills and website building experience to design and build www.ebikebook.com The mission of that site is: "Build a Community that Promotes Electric Bicycles as a Green Transportation Opportunity." I have many fun ideas for that web site. Starting with the community, the website will point to a www.facebook.com fan page (or similar tool, if facebook.com does not allow use of my URL since

www.ebikebook.com is similar to www.facebook.com)
I sure hope that facebook.com does not block my
domain. As the community builds I hope that it
becomes an unbiased and a premier source of user
review and feedback on conversion kits types, and even
integrated bikes. Users can share pictures, ask
questions, help each other, and communicate within
this rapidly evolving community.

*What is the width of the rear wheel mount at the axle
line?*

I like my front wheel mounted e-bike motor. It
corners like a motorcycle, the extra 12 pounds (of a 600
watt hub motor) helps with front wheel traction. It more
evenly distributes that additional 40 pounds across the
bike as opposed to the alternative of a rear axle
mounted motor. It helped me avoid many installation
issues. For example the rear wheel on a mountain bike
would have 6 or 7 new gears to test and adjust the
derailleur alignment. What an oily mess that would be
working with the chain.

There are positives to having the weight on the rear
wheel. For example, when you stand the e-bike up on its
rear wheel to go over steps or curbs it will require less
strength. Also, if your front forks do not give you the 3
and 3.5 inches width needed at the two measuring
points, then you have this option to consider.

What is the width of the rear wheel mount at 5 inches above axle line?

This is related to the same as discussed above for clearance 5 inches above the axle line so that you have enough space for the motor. The hub motor in both front and back locations requires a clearance of 3.5 inches or more from the axle line to everywhere within 5 inches away from the axle line.

RIDES:

Where do you plan to ride?

There are 18 questions in this section. This answer to the above question should be given in terms of specific named rides and destinations. For me work is too far and there are no bike trails to get there. I don't plan to take this bike with me on local vacations. I plan to ride mostly on weekends or days off from work to reduce the number of miles that I drive my car by several hundred miles annually. I want to ride the e-bike instead of drive my car for any local errand or social event within three miles distance with weather, time, and trip scope permitting.

Do you see what I did? I eliminated the "Commuter" and "Foldy" designs and isolated my needs. This is a high level summary of what you what to accomplish. Please be as specific as possible. You should name specific routes, distances, and destinations if you will frequent them. The answers to the next 17 questions should be in line with the answer you give to the above question. Also I must say that in terms of issues like "time permitting the use of the e-bike" the bank trip I took yesterday to a crowded bank parking lot, and crowded shopping center...left me convinced that I went faster by bike than by car on that particular trip!

Will you be riding alone or with others?

When cycling with others, compatibility in speed and distance is critical for all to have fun. If you plan to ride with others, you will want a system that is at least as powerful and long lasting as theirs. The alternative would not be fun for you or them.

I imagine in the near future groups of e-bikers, sticking together for a safer commute to work. I bet there are already a few groups like this perhaps in a place like New York City. I have not seen one yet here in LA, but I'm sure it won't be long before I see the first one. I think that like websites that promote carpooling today will pop up to promote "ebikepooling" for groups of e-bikers to commute together for safety and camaraderie. Hmmm...that's a good idea. Excuse me I need to register a new domain!!!

I got it!! www.ebikepool.com is mine and part of my ebikebook.com mission to help establish an e-bike community on the web to help promote and take advantage of the e-bike as a green transportation opportunity. I guess once this book is out, I will study several carpool sites and come up with a way for local e-bike commuters to learn about each other, share information, and work together for safer commuting.

If planning to use the e-bike with others, then do they have e-bikes too? (What kind and what power do they have? And how often will you be riding with others?

Keep in mind that the more alike your e-bikes are, then the easier for you all to stick together. The better bikers stick together on the road, the easier it is for cars to see you. Ultimately the safer you will be on the roads.

Are there safe bike lanes?

If so, then you are lucky! E-bikes will help you take advantage of the bike lanes in your neighborhood, and reduce your miles driven in a car. While I do have fantastic beach and creek trails here in the Santa Monica Bay only 2 miles away. The major roads are often hazardous for bikes. So I often jump onto sidewalks and focus my routes primarily on the minor neighborhood roads that are almost totally free of moving cars.

Will you be riding before dawn or after dusk?

For some people the majority of their miles ridden will be in the dark. Commuters will make up most of the "midnight ridazz" or predawn riders. In my research for the perfect kit to meet my needs and budget, I ran across a couple of kits that also included an integrated light to the battery. The light looked significant, 3 relatively large halogen bulbs for the front light. If you are planning to regularly ride in the dark at higher speeds on an e-bike, then for safety alone you may want to consider a conversion kit that addresses the lighting issue. If you want to ride with a group at night in LA check out www.midnightridazz.com . Or for LA's self proclaimed largest monthly ride check out www.lacriticalmass.org .

Are there side streets, bike paths, sidewalks that you can utilize for most of your ride?

If no, and you live on an interstate highway, then ask yourself how do you feel about riding a 20 mph vehicle on that highway. Will it be safe when big trucks go whizzing by. The main point here is to look at your primary uses and not just the distance that you would drive in a car on the major roads directly. Think about the distance you would ride. For example, my cycling route to the majority of the markets that I e-bike to is about half a mile longer in each direction than if I would drive a car.

How far is your average ride?

Think about it in terms of miles and then compare it to what the conversion kits are claiming. Be conservative. If the distributor or manufacturer is stating up to 20 miles, then cut that number in half or even 1/3. So if you need to cover 10 miles in one stretch, then I would power up to with a battery and motor configuration rated for at least "up to 20 or 30 miles". One reason to be conservative is that batteries do get old and hopefully wear out slowly. You'll see your high power green light stay on less and less over time. Or you'll see a drop from high power green light to lower power yellow light indicator earlier on the big hills over time.

The average distance of the ride is relevant to the type of design you select. Overall and by definition the style of bike selected will dictate the maximum range, other things being equal. For example, when I use different types of my regular bikes I can go significantly different speeds and distances with the same level of exertion. Here are the six styles listed in order of their ability to handle distance with the longest range style listed first: Touring, Commuter, Local Errand, Combo, Recreation, and Foldy. Of course there are always exceptions, but this is generally the correct order.

Are there hills?

Big hills and weight (cargo and rider) are an extra drain on the battery. You'll want to just go half throttle on the bigger hills and work your legs to get up and over the hill. Sure you can go full throttle on the hills. I still do often, but not if I have to go a total of 5-6 miles. I like to have a little left to make the last hills to my house. I try not to exceed 6 miles round trip with my e-bike that is rated for "up to 20 miles". I'm 200 pounds, and with the baskets, lock, waters, pump, rack, 3 baskets, groceries and e-bike conversion, I bet that my e-bike and I together weigh just under 300 pounds. I guess that is a heavy load for any battery and there are lots of hills in my neighborhood. Round trip, 5-6 miles is about the limit with a 2-4 large bags of groceries.

Is there wind?

Is there a prevailing wind direction, which you will consistently hit on one of your frequent routes? Will there be a head wind or tailwind? If there is a 10mph or higher headwind for a significant part of the way, then be conservative and cut the "up to 20 miles" rating by 75%. Plan on getting 5-6 miles out of an SLA battery if you, your cargo, and your bike weigh together about 300 pounds. Same as if you had hills. And if you have hills *and* significant headwinds cut it even more. Maybe you get only 3-4 miles in a hilly and windy place with a battery rated for "up to 20 miles."

What is the maximum load of items like groceries that you want to carry? (weight and number of traditional brown paper grocery bags)

Are you shopping for a family of 5 like I am doing? Will the age and appetite of your kids impact the miles and locations that you will be riding to? For me I like to hit a couple of stores. I'll start at Trader Joe's for 2-3 large brown bags, and then hit the farmers market on the way back for some extra fresh greens, veggies, and fruit. I have a web bungee for all my baskets. I'll talk more about the web bungee later in the "Using Your e-Bike" section (aka Part Four).

Do you want to ride to work?

Look at the practicality of the time of days that you work, your average weather, and the time of day that you have to start work. So you may want to ride an e-bike to work, but is it realistic? Would it impact your career negatively? Is that important to you?

How many miles to work?

The cycling route to work may need to be longer, and the speed will most likely be slower. Do you have the additional time in the morning and evening? Will the e-bike conversion kit in your budget range give you the distance needed, conservatively as described above?

How safe would the ride be for a 20-25 mph e-bike?

Safety and risk avoidance is very important to all that read this book. Where I live there are many roads that would not be safe to habitually ride to work by myself. My commute to work is 17 miles away by freeway and on an e-bike cutting across the city during rush hours for that distance is just not safe or practical, even if the extra time wasn't a factor.

What is the traffic like at rush hour in the AM and PM?

Perhaps ride a regular bike one day without the e-bike motor on that route at rush hour. How did you feel? How do you think you would feel if you were going 5-10 miles per hour faster on average and with less effort? Pay attention to the afternoon commute going home. How safe is the commute? Remember over 1/5 of the cycling fatalities occur in the 3 hours starting at 3pm. This 3 hour window has the highest percentage of fatalities in the day.

What is the road condition? Pot holes? Gravel? Dirt/Mud?

Riding on less than perfect asphalt streets is another good reason to use a mountain bike frame with front and rear suspension as your e-bike. Since you are riding at slightly faster speeds with an e-bike, you will feel the bumps more. The tire sizes and tread patterns can be optimized for the type of terrain you'll be riding. I switched to a much smoother pattern than my previous knobby tread pattern due to the higher speed of an e-bike. I knew the wind resistance would be worse if I left the knobby tires given the higher e-bike speed.

Have you spoken to your boss about you considering riding an e-bike to work? Are you the boss?

If you are the boss, then this should not be an issue unless you need a car at you place of business for other issues. An e-bike for the purpose of commuting is a significant investment. For this reason and many others I highly encourage that you speak to your boss prior to you purchasing and/or building a "commuter e-bike".

What does management think about your e-bike parked and charging in the warehouse or your shipping/receiving area?

Every e-bike commuter that I have spoken to charges their e-bike at work and parks it indoors. They were easily able to get the OK from their employers to do this. However, there are going to be work environments where this is not OK. Ask first, or be sorry later.

If you are planning to commute to work on your e-bike, you need to consider where you will park and charge it. Your ability to park inside or not will be a very important aspect of the bike's configuration. If you have to park outside, then bike theft is an issue to worry about. I discuss bike theft more in the Environments section in a few pages forward.

Then, the other issues revolve around the battery. First, how are you going to charge it for the ride home? And secondly, how are you going to secure the battery to prevent theft?

If you have to park outside then I would highly encourage selecting a conversion kit with an *easily removable* lithium battery at 6 pounds. Not only does it prevent theft, but you could charge the battery at your desk or work area during the day for the drive home. There are some nice kits out there that have lithium or SLA batteries that quick snap in to the integrated rack with a key lock for the battery, and/or have a carry handle integrated to the battery.

SLA Battery bags that Velcro attach to the rack are not as good a fit, especially if you have to remove and attach multiple times a day. I just used 3 tough 24 inch industrial zip ties to attach that 20+ pound battery semi permanently and very snugly to the rear rack. I never remove the battery.

If you have to park outside at work, you'll be risking theft. You'll likely want a removable lightweight lithium battery when you arrive at work, for charging at the office, and then snapping in the fully charged battery for the ride home. If work is concerned about the electricity cost remind them that multiple sources have estimated the cost per mile of electricity to be about 2 cents per mile and that you can pay them $1 per week if they like.

Do you plan on "Touring" with your e-bike?

Touring with an e-bike is something of an exciting concept for me and didn't even think it applied until I finished my first draft of this book. It was only when I started searching the Library of Congress – Subject Headings, that I realized my mistake. Bicycle Touring and Touring books are very popular globally. I apologize to all the Touring Enthusiasts for my lack of touring experience. It is something that I have never done, but of course dream about often. In fact, I see the opportunity to Tour with e-bikes as a huge opportunity for growth in the Cycle Touring industry globally. So much so that I will go and attempt to register www.ebiketouring.com right now. Though probably too late. WOW!!! I got the domain!! AR Publishing Company now owns this domain.

This is the one major type of bike that I do not own. The next e-bike that I build in a few years likely will be on a touring frame. The attributes of the design of course will be the ability to go 50 miles or more pedal assisted. Ideally the Touring bike would have neither a thumb throttle nor a grip throttle. The power would come from pedal pressure assisted throttle, because if you are in the saddle for 3 hours a day there is no way you would want to have your thumb hold a thumb throttle. Though, you might be OK with a grip throttle. I'm guessing a nice configuration would be a 48 volt system powered by a lithium ion battery and driving at least a 1000 watt motor. Can't wait to build that one!! Or the website that will allow a community to develop and evolve for touring cyclists that use e-bikes, host e-bike touring events, or want to consider e-bikes for touring. The main point here is that touring is tough work especially in the hills, unless you are built like Lance Armstrong. E-bike technology will open up Touring to a significantly larger pool of the cycling population; this thought alone is very inspiring.

I hope this prediction rings true, and one day I get to enjoy a vacation with my wife on an e-bike assisted touring trip for a week through Napa Valley or some other Californian Wine Country. My wife likes to cycle, but not as much as me. She likely would not try Touring even though she is a tough Sierra Club leader and can beat me up on the trails. An e-bike environment for touring would make a significant difference in her agreeing to try a bike touring vacation. More to follow on the web site www.ebiketouring.com by the Fall of 2011 this site should be in full swing.

ENVIRONMENTS:

Where will you store your e-bike? Indoors?

An e-bike needs to be stored indoors in a secure garage with easy access to roll in and out of your garage. Excessive heat or cold can damage the battery. Another good reason to have a removable battery is that on the hottest and coldest days/weeks/months of the year you can store the battery inside your dwelling

Since I semi-permanently attached my SLA battery to the rack with zip ties, I needed to do something extra to temperature control the garage. I live in LA about 3 miles from the beach, so cold was not really a concern. But it can get hot, and the closed space of the garage can really heat up on days over 90 degrees, not to mention the days of 100 degrees. To minimize this issue of too much heat in my garage, I installed a large thermostat controlled attic fan on the large vent at the high point in my garage. I also have a couple of these attic fans in my attic, they are great and highly encouraged for most homes and priced on-line at $80 or less. The last one I bought was only $50. Also good to know is that an overheated battery does not accept a charge well.

Is the bike storage location easy access? Any stairs?

Remember that an e-bike can weigh as much as triple the weight of the bike before the addition of the rack, baskets, motor, cables, batteries, and power inverter, especially if you are using SLA batteries. For perspective, I am a 200 pound weekend warrior and I feel that getting my e-bike up two steps or stairs is doable but no fun. Four or more steps or stairs is not really practical to attempt by myself given that my bike weight has tripled.

Also related to the extra weight of the bike and storage is this next end result. I used to hang this mountain bike easily on a rope and pulley, bike hanging device in my garage. Well you may have noticed the past tense verb in the preceding sentence. The end result is that I can no longer hang the bike with the rope and pulley system due to the new weight of the e-bike. Perhaps if the battery was easily removable it would still work for the e-bike.

Is it secure?

Although crime across L.A. is dropping, bicycle thefts, which rose 29% last year, is the glaring exception. Nearly 2,000 bikes were reported stolen last year. Authorities believe the actual number of thefts was much higher because so many people don't report stolen bikes. (Winton, 2010)

The top ten (really 11) cities for bicycle theft according to the U.S. market leading bicycle lock manufacturer Kryptonite: (T., 2007)

1. New York City
2. Chicago
3. Boston
4. Philadelphia
5. San Jose

6. Los Angeles - tie
   San Francisco - tie
8. Seattle
9. San Diego - tie
   Washington, DC - tie
   Portland, Oregon – tie

Tips from the Kryptonite blog to help maximize security of your e-bike:
- always lock your bike, especially when at home
- two types of locks used at the same time are better than one
- lock to a fixed, immovable object that cannot easily be cut, broken or removed
- lock in a visible, well-lit area
- lock in a location where there are other bikes
- if you commute, change up your locking routine so there isn't a pattern – lock one place one day and somewhere else the next
(T., 2007)

Can your kids or neighbors' kids get to it in one of the storage locations?

This is obviously important for the health and safety of others, perhaps yourself too. If a minor takes your bike for a joy ride with or even without your permission, you could be liable for negligence both financially and criminally. So, what are we to do?

I suggest five things to do, if kids are around to prevent this potential liability:

1. Do not allow minors to ride your e-bike.
2. Acquire a heavy duty lock and always use it.
3. Lock up your bike at home, when garaged.
4. Select a conversion kit with a key lock start and/or a removable battery
5. Don't park your bike in an area where unsupervised kids have access to the bike.

Do you plan to load the e- bike into a truck/van or onto a car bike rack?

If you can't dead lift an eighty pound Olympic straight bar and push it over your head a few times, then you may want to forget about loading the e-bike into a car or bike rack the way you used to load your regular bike *by yourself*. Now it's a two person job. In fact when loading by e-bike with the SLA battery in my cargo van, I like to use 3 people to load it. One person goes in the van and one person on each side of the e-bike, which you can imagine is significantly easier and safer for all when loading the e-bike.

What is the average weather like?

You'll want to look at this from several perspectives, and especially these two. How many months have average high temps of 60 degrees or more? These are your pleasant temperature months. How many months have ice and snow? These are the months that the e-bike will likely get dusty in the garage, and the battery will sit in the house somewhere so as not to get damaged from the extreme cold in a garage that is not heated.

Though having said that, I bet one day "Nanook of the North" (whom I respect greatly) will read this book and dream up how to build a 3 wheeled, spiked tire, frozen lake ice bike for getting out to the ice fishing hut faster while hauling the fishing gear. Nanook would need to use a removable Lithium battery and to keep it from the cold while in the hut fishing. A Nanook knows below freezing temps can quickly rob a battery of a significant portion of its power.

Do you plan to ever ride in the rain?

There are a lot of issues here. For starters, 36 volts is about 1/3 of the current that runs through your house outlet. It could give you a shock that you would not quickly forget. If you are in a rainy city like Seattle or Portland all is not lost. If you plan to ride in the rain, the good news is that there are kits that rate the motor / hubs as sealed and water resistant.

Will your bike sit idle for several months at a time?

If the bike will sit idle for months, then you should think about where you will store the bike. Perhaps use a plastic or cloth to cover the e-bike and keep the dust and dirt off. How cold does it get in your garage or storage area during the time of extended non-use? To prevent damage to the battery, or just help the battery last as long and as strong as possible, it is always suggested to store the battery indoors in your dwelling or area that is temperature controlled.

Is there a place indoors to secure, and charge your bike at work?

The ideal scenario for any cyclist or e-biker commuter is to be able to park their bike indoors. For the e-bike commuter this means that once you get to work you can park and charge your e-bike. A warehouse usually makes the perfect location. If your place of work has a shipping and receiving dock, warehouse, or area you can roll right up and plug in, then it is time to talk to your boss about your new e-bike parking spot.

Other Issues?

Think of the "Other Issues" in terms of the above Five Areas of the Usage Needs Survey. Those are again:
Goals:
Rider(s):
Bikes:

Rides:

Environment:

Spend time reviewing each of the questions on the list one last time to uncover any other hidden needs that may impact your choice of the best fitting conversion kit. I remain convinced that the most important part of solving a complex problem with new technology, revolves around a thorough and detailed written needs analysis.

The needs analysis will help keep you on target to save you time and money. It also allows you to best set your own expectations, as well as the expectations of your friends, family, co-workers and employers. Best of luck picking the best fitting conversion kit!

•

Finally, you are at the point where you can feel better informed and answer with confidence the 3 questions that I presented at the beginning.

The three questions again are:

1. Should I purchase e-bike technology now?
2. Should I buy an integrated, finished e-bike that is ready to use, or should I buy a conversion kit to upgrade a bike?
3. What power and bike style should I use?

These are not easy questions to answer since there are at least several hundred dollars and even potentially thousands of dollars involved. No one wants to make a big, foolish mistake on this decision. And that is why I wrote this book. I felt that the information I wanted to make a more educated decision was not there. A book like this one was not available. There was no on-line community of e-bikers that was free and loaded with endless and current user generated photos, reviews, and blogs about e-bike conversion kits. I felt I went into this e-bike decision with barely one eye open, and got lucky with my all my decisions on building my first e-bike.

Though I still am no expert in all brands and styles, to do over I would likely get the $699 Wilderness Kit, same motor but has a removable lithium battery. I feel that after 8 months of weekend riding that the SLA battery is at less than half its initial capabilities in terms of torque and distance. At this point I'm working with www.falconev.com to build a customer battery pack that will likely greatly increase my distance and satisfaction. Look for news on this project at www.ebikebook.com .

•

I hope you get the opportunity to visit prior to your purchase www.ebikebook.com . My sincerest hope is to create an on-line community of e-bikers that share pictures, experiences, resources, reviews, events, and ultimately work together to promote this new *green transportation opportunity*. I hope that you will share your results and even photos in this free community.

Part Three

Installing the Kit

After many hours and days of work analyzing my needs, exploring alternative kits, researching manufacturing/distributor options, and decision making... the moment of opening the conversion kit has arrived! Still some questions exist for which I have no answers, but that's okay as this is my first e-bike and I expect to learn by doing. There are so many options and issues that I bet many potential e-bikers get "paralysis by analysis" and that is they never just go for it and build the bike. In fact there still are questions I have about the technology that remain unanswered and may remain so. For example how many miles will the motor last? How many miles or months will the battery last? Will the distance traveled today on a single charge be greatly reduced in 6 months when the battery is older? The answers to these questions will depend on many factors, and over time I hope to learn the answers.

Ultimately, I decided that if I just go for it, then I will learn about the technology while I use it. The point here is to just relax and enjoy the e-bike building process even though there might be surprises or missing information. I made some decisions based on hunches and hearsay, like the motor wattage and battery volts. The education you get from becoming a user of the e-bike technology will quickly add up as you get hands on.

Here are the Installation Steps that I went through plus a few extra you may want to consider:
(pictures are found in the paperback book, or at www.ebikebook.com)

Preparing your work area and tools –

For me, a wide open area in the garage worked best. The very first thing I did in the work area was layout a drop cloth since I was planning to clean the chain too. I like to turn on a local radio station to keep me fired up on long projects. Also, a friend told me that if you have a Bud Light or two handy that the installation project will be less tedious as well.

The tools used were a screw driver, an extra long Allen wrench set, a couple of wrenches for the wheel nuts, a large bag of extra zip ties, a pair of wire snips or scissors to cut the zip ties/rim tape, a bike tire removal tool, and a second pair of hands for a few minutes.

Layout your bike on its side, your tools, and boxed kit all within easy reach. Make sure you have good lighting in your work area as you will likely be there for a few hours. You may need the help of a second pair of hands for a few minutes to stretch the front forks up to a half an inch to fit the likely slightly wider new front motorized wheel.

Don't underestimate the importance of mentally preparing yourself for the surprises including failures that may happen. I failed on the first e-bike installation (that $65 www.craigslist.com bike) due to lack of clearance at the forks. I did not have the needed 3.5 inches width clearance at 5 inches out from the axle line. Luckily I had a "Plan B" bike. Then I ran into some extra work and grinding to use some safety latches for the front axles that did not fit the slightly bigger diameter axle bolt and that added about an hour. I spent about 5-6 hours from start to finish including a break I took to have quick 20 minute dinner. When I was done, everything was perfect to me, including the many zip ties to hold the wires.

Relax and enjoy the journey that is the process of building and creating, as it may take you more than one day. In other words if you are a "Type A Personality" like many of us, don't stress out if it takes you longer than you thought it would to complete. Avoid setting expectations with others or yourself as to when your inaugural ride will occur. It will happen at a time when it is ready and not before.

Preparing your bike –

For me this step took about an hour. Having the drop cloth was definitely a good thing. For the past 20 years,

I have been using silicone to clean my bikes and motorcycles with great results. You'll go through about ½ a roll of paper towels from start to finish. You will want a nearly full spray bottle of silicone, and a used/clean toothbrush. I wear latex gloves for this part and go through two pairs of gloves changing after I finish cleaning the chain / derailleur. Start the cleaning process with the derailleur then move on to the gears brushing the teeth of the gears like the teeth in your mouth. Then use the brush and lots of silicone and paper towels till you feel the chain looks reasonably clean. This will splatter spots of grime on your frame and rear wheel, which is the reason you always start with the chain and gears first. Certainly there are other ways to clean the chain, just Google bike chain cleaning. But this way has worked for me my whole life and I never had to replace a chain!

Now change latex gloves before moving on to the rest of the bike. Best to leave the garage door open while doing the cleaning for ventilation. The good news is that silicon spray doesn't smell nearly as bad as WD-40. You should spray silicon on literally everything and then wipe it off. I typically leave the rear wheel areas as the last item, so I don't soil my second pair of gloves too early. Test a small area of the seat, as you may not want to spray silicone on the seat, though silicon is fine on vinyl seats.

Now, open the front brakes so the wheel can be removed, remove the front wheel, and remove the right handlebar grip. What is the condition of the handlebar grips? It would be a good time to replace if necessary. Also, it is a great time to add a "Rear Wheel Kick Stand". Install new rear wheel kick stand –

Installing a rear kick stand is an excellent idea with many benefits. For example, you can now load your groceries into the rear baskets with ease and confidence that the bike will less likely fall over. I found this kickstand on my favorite www.amazon.com that there are many types of rear kick stands. I found the one for mountain bikes that was the heaviest duty, had good reviews and was $15. It is called the Greenfield Stabilizer rear mount kickstand. It still amazes me that the kickstand supports the weight of the bike with the baskets loaded and that 20 pound SLA battery mounted to the rear rack.

Opening the kit box –

Now that your bike is cleaned, prepped, and has a rear wheel kick stand you are ready to proceed. Use a key to cut the thin Chinese tape on the box. I avoided using a knife to open the box as I didn't want to inadvertently damage any of the many wires in the box. Remove all the contents, then inspect for damage and inventory items. There was a one page sheet with some instructions and inventory list. I looked at it once, and set it aside. That installation sheet was surprisingly short and one of the reasons I felt the need to write this book. It was more of a packing list, than an installation guide. I didn't really use it. I imagine most customers in the US felt the same way about the one page sheet.

Rim tape –

A very important point, one of many overlooked by the installation sheet included with my kit, was the need for rim tape. Rim tape is cheap and available in many styles, and most types work great. Without it, I understand that there is a much greater probability of popping a tube. I like the rim tape with an adhesive backing and a cloth front. I purchased my latest rim tape on www.amazon.com for $9. It is called Zefal Bicycle Rim Tape (17mm).

Thorn resistant inner tubes –

This is the only way to go for all of my bikes. These tubes are a little heavier, but well worth it compared to the alternative of being stranded attempting to fix a roadside flat. I still like to carry an extra regular tube on most of my bikes since they are light weight. On www.amazon.com I purchased the Avenir Thorn-Resistant Schrader valve MTB tube for under $10. Make sure you put rim tape on the new motorized wheel as most of the Chinese makers of these kits do not include rim tape. This is a popular gripe in reviews on www.amazon.com of bikes and wheels from China.

New tires –

I am really happy with the tires I have. They are fat for bump absorption (to protect rider and wheel) and relatively smooth on the surface. They corner on the city streets quickly and smoothly. These almost smooth tires cut through headwinds with much less wind resistance than the typical mountain bike tires. Once I wear them out I will replace them with a tire that has a Kevlar belt for better protection of the tube and improved reliability.

Install the wheel –

This is one of the most exciting steps. When you are done the bike can stand up on its own with the help of the new rear mount kick stand. Take care to follow the arrow direction on the motor and install the motorized wheel so that the wheel turns in the correct direction!!! I needed a second pair of hands to fit the new motorized front wheel on my old Diamondback mountain bike forks. My younger son was there to oblige. Like trying to break a roasted chicken wish bone for good luck, we pulled and stretched the forks until we finally got the new wheel to fit. Carefully check that the wheel is completely in the forks before and during the tightening of the front axle nut.

Reconnect and TEST THE BRAKES –

The first thing you will notice is that the rim is most likely wider at the point of contact with the brakes. So I had to let out almost an inch in the cable to buy me the one inch in extra rim width. Then adjust the angle and position of the brakes. Be sure to inspect for perfect contact of brake pads to wheel. Once positioned correctly, give them your strongest squeeze to make sure you tightened the cable properly. Test the brakes before you head down the road. Check them for rubbing and for adequate stopping capabilities. To check for rubbing lift the bike with one hand and spin the wheel looking and listening. If the brake rubs and touches only on one side, adjust the side in question typically by unhooking the spring arm and them bending it outward so that it pulls the touching brake away from the wheel. To see this done just search www.youtube.com for "adjusting mountain bike brakes." I will eventually put a link to a video on this on www.ebikebook.com as well.

Install kit rack, use existing rack, or buy other rack –

I already had an excellent rack on that bike with two folding baskets installed on the rear rack. It's a tough Blackburn rack and plenty strong for my intended use. If you are going to put rear racks on a bike, you should find a beefy aluminum rack. The longer the rack the better as you will want the rear rack baskets as far from your pedals and heals as possible. I already lost a bolt that attached the rack to my frame while on the road. If that happens again to me, then I would consider using a product like Loc-Tite to keep the bolt in place. Carry an extra bolt or two in your bike bag under your seat.

Install baskets and bungees –

I'm very satisfied with the products I selected for my "local errand" style bike. The internet is a fantastic resource for bike parts and you have many options. Start with www.Amazon.com and consider investing for one year in a Prime Account status so that most everything you buy ships with 2 day air service and no charge shipping. My Prime Status pays for itself annually several times over.

I installed two Sunlite Folding Rear Rack Baskets – Black Color costing under $15 each. They need to be installed as far back on the rack as possible to give your hard working feet the needed room to prevent contact. In fact try to get an extra long rack so you can mount the rear side baskets even farther back to prevent your heels ever touching the baskets while pedaling which could be an injury or worse.

I installed the Wald 135 Deep Sized Grocery Front Handlebar Bike Basket costing under $14. This is the largest basket on the bike, and carries weight very nicely. If you pick up at least two of the Sunlite Bicycle Bungee Cargo Net Black costing under $13 each you definitely will not be sorry. One is used for the front basket and the other for the rear. You may want to carry a third bungee web for strapping another item on top on the rear baskets should you like to stop by garage sales on the way home like I often do hunting for bargains.

Install the inverter box –

The inverter box was relatively simple to install, once I decided where to put it. I tried several scenarios holding the cables to make sure I had sufficient length to reach the front hub motor before the first zip tie went on. One of the potential spots did not work due to cable length. Also think about accessibility to things like the inverter on/off switch, power indicator light visibility, and even cooling. The inverter does get a little warm. Look at the pictures of where I installed mine. I secured it to the front of the left folding basket. You will use a bunch of zip ties to secure the inverter and cables. If you mirrored my installation approach, then you will see that the folding basket will need to remain open. Once the inverter is secured connect the inverter cable to the coming out of the front hub motor.

Install the throttle –

The thumb throttle is included with the Wilderness Kit. Installation of the throttle is straight forward. I left the brake levers and shifter switches alone. I removed the grip and then cut about 1 inch off the grip section that first fits on the handle bar to make place for the new thumb throttle. See the pictures in the picture section of the book. Or see color pictures at my web site www.ebikebook.com .

Installing the battery –

The 36 volt SLA battery bag from the Wilderness Kit comes with three 12 volt batteries prewired together in a zippered battery bag that has 4 Velcro straps on the bottom of the bag for connecting the battery bag to the rear bike rack, see pictures. You will then connect the battery cable to the inverter. See below the industrial zip ties I used to more securely attach the battery.

Zip tie loose wires and battery –

You'll want two types of zip ties. I used the 8 inch black ties for the cables, inverter, even the racks. I used some clear ties for the wire that goes underneath the top frame tube from the front forks to the rear rack. Then, I used three 24 inch zip ties to very tightly strap the batteries and battery bag to the rack and stop the 20 pounds of batteries from bouncing off the rack when I hit big bumps at speed. The Boston Industrial UV 175 Tensile Strength 24 Inch Cable Ties worked fantastic for tightly securing the battery bag to the rack. That 24 inch Cable Tie will also greatly reduce the risk of electric bicycle battery theft.

I feel the following is an accurate long range crime forecast. I bet that in ten years that e-bike battery theft will be easy money for the bike thieves. A lithium battery is only 6 pounds today and fits in a backpack. It can sell for more than the average stolen bicycle.

I probably used 30 zip ties across the entire kit installation and another 30 zip ties to install the two folding rear baskets. I used wire snips for cutting the excess zip ties, but scissors work well also. Be careful with both to not cut any wires by mistake.

Install the zip ties in a way that will not cause personal injury. It is very easy to get a nasty wound from a zip tie on your leg. Make sure to do two things with every zip tie. First, cut the excess as close as possible, meaning the most possible excess strap has been removed. This leaves a little piece sticking out with sharp edges that can easily cut a fast moving leg. So always position each zip tie in such a way as to minimize the risk of injury.

Water bottle and bottle cage –

I replaced one of my two cages which was about 20 years old and fully depreciated. In other words, the cage was kind of flimsy. I was concerned about the water bottle cage losing the bottle. Since e-bikes mean your average speed will be a little faster you will be hitting the bumps faster and harder. Your water bottle should fit snugly and securely. If it is loose, bend the cage a little inward with the bottle out to create a snugger fit.

Speedometer installation –

This is optional and assumes that you do not already have one. Speedometers (aka odometers) are beneficial so you can gauge your speed during regular sections of your ride, and keep an eye on the long term performance of your battery performance in terms of distance. You can keep track of your total miles by month bragging about your mileage results at www.ebikebook.com. The one that I chose is the SIGMA BC906, 9-Function Topline Wired Bicycle Speedometer it was $28 on www.Amazon.com . The directions are straight forward and detailed, follow them step by step. Perhaps prior to purchase and installation check out videos on how to install a bike speedometer on www.youtube.com .

Lights –

Are you riding predawn or post dusk? Remember you will be riding faster so being seen is not the only issue. With the extra weight your maneuverability is less, and braking distance is longer. Meaning a strong light is needed to see road hazards earlier. Think safety first and err on the side of too much light if you are a regular to riding in the dark. The lights that I purchased for cheap ($20 total?) on www.amazon.com are the Planet Bike Blinky "3" 3-Led Rear Bicycle Light for under $10, and the Planet Bike 3042 Spok LED Micro Headlight.

The very good news is that these lights contain the battery that powers them. All you need to do is locate a good place to attach them to the bike. They are very small and I carry them in my under seat bike bag for use only as needed. They attach in seconds to the handlebar for the front light, and then the seat bag for the rear red blinking light.

Horn and Bell –

If you plan to ride in areas with pedestrians like shopping center parking lots, then this is a good idea. Remember that you will be going faster than regular bikes and quietly. Pedestrians likely have never seen an e-bike, and will be surprised at your speed while not pedaling. It's counter-intuitive to them. And yes for many reasons you should have a bell, including preventing civil liabilities of vehicular caused injuries. You need to do your "due diligence and ordinary care" to do everything possible to prevent injuries to others. This will help lower the risk of being found *negligent* should you ever hit someone with you e-bike and get sued. If you have a positive net worth, buy a bell, such as the Incredibell. Then, talk to your insurance agent.

While the Incredibell is an excellent and polite way to announce yourself to pedestrians unaware that you are approaching, it is not a good safety device for getting the attention of a distracted driver. When biking in traffic, it is easy for a driver pulling out into the road to miss a speeding bicycle. Since I cruise my e-bike on major LA roads occasionally, I purchased a 105 decibel compact bike horn for only $20. It simply attaches to the handlebars and the battery is self contained. The button is placed where your thumb can get to it quickly. The reviews were very positive on Amazon and suggested exactly what I wrote. As you can imagine if you blast a horn like that at a pedestrian you might get a shoe thrown at the back of your head. So, for me I will ride with both a bell and a horn. The loud horn is for the cars and the bell is for the pedestrians.

Tools, portable air pump, and patch kit –
I have a patch kit and pump with all my bikes. I include the tools, if any are needed to remove the wheel from the bike and tire from the wheel. Ideally carry any tools needed to secure your bike rack if it gets loose. Especially important is to carry an Allen wrench to tighten any loose frame bolts that attached the racks or water bottle cages. Carry an extra bungee and some zip ties in case a basket or rack comes loose as well.

In fact, my rack came loose just last weekend, on the way home with a dozen doughnuts for my skinny teenagers (ok and me too). I lost the right screw that holds the bottom of the rear rack to the frame. With a 20 pound battery, the inverter, and two baskets attached to the rear rack, I could not go further. Luckily I had my MacGyver hat on and was under way in a few minutes. First, I checked my tool bag. Darn, no Allen wrench! Wish I had a written and read my book prior to using my e-bike! If I had the Allen wrench I could have removed one bolt from the water bottle cage and used it to secure the whole right side of the rack.   What I did have was an "old school bungee cord," which worked perfectly. I just straightened out the plastic coated wire end hook a little and fed it through the rack and frame, then bent it to a closed position. I had it fixed before I could finish humming the MacGyver TV Show theme song.

Reusable Shopping Bags –
    I have two bags that I keep in the rear open basket. I do not use a reusable bag for the large front basket which I use to store my helmet and gloves when the bike is not in use. Trader Joes had the perfect reusable bags for 99 cents each, which fit perfectly in the basket. I use Trader Joe's paper bags basically as a liner that I throw away occasionally to keep the reusable bag sanitary as I use it for hauling food for my family.

U-Lock, and Combo Lock Cable

Of course the more expensive the lock, the better and more secure you should expect it to likely be. Kryptonite has always made some great U-locks that are somewhat bolt cutter proof. I purchased mine on www.amazon.com the Kryptonite Kryptolok Series 2 ATB Bicycle U-Lock (5-Inch x 9-Inch) which works great and is around $20. And for most of my bikes I have also purchased at www.amazon.com the Avenir Coil Combo Cable Locks (10mm / 6 feet) for under $12.

Insurance –

Of course there are several types of insurance to consider. Comprehensive insurance for theft or damage is probably too expensive and not worth it. But if it is cheap enough I might get it one day when I build a more expensive bike. I have medical insurance for myself, but there also is medical liability insurance should you crash into and hurt someone, just like you likely have on your car today. Then of course, property damage liability insurance is available. Of all the liabilities that are possible the one that scares me enough to actually purchase insurance against is medical liability insurance. And it surprises me that after 35+ years of crashing into things on my bike that just recently was the first time I ever thought of covering that liability. With the extra weight of an e-bike and additional speed of the bike, the risk of causing a serious injury to someone is greatly increased in my opinion. Check out www.ebikebook.com for ideas and suggestions regarding protecting your assets.

Part Four

Time to Enjoy!!

Whew! OK I think I installed the e-bike kit correctly and now the battery is charged. The test drive confirmed this as I flew in the dark up and down my street to the amazement of myself and my family members. Mission accomplished at 11pm, it was time to lock up the garage and relax a few minutes as I reflect on the past 6 hours of bike cleaning, maintenance, and e-bike conversion kit installation.

I woke up the next day feeling like a kid at Christmas! I couldn't wait to get through breakfast and head out on my first few "local errands". I went out to a local Trader Joe's Market for a couple of large bags of groceries. Those fit perfectly in the rear baskets, and then held in place with the web bungee.

Next, I hit the local Farmers Market. I'm able to park 3 feet from the front door of the market, and I can see the bike the whole time as I am never more then 20-30 feet away. I even just leave my helmet on since it is such a quick and convenient stop. Since I am so close to my bike, it's no problem leaving goods from the previous market stop in the rear baskets covered with the web bungee. I pick up a large bag of fresh greens, veggies, and fruits. I put the third bag in my front basket and then cover it with a second web bungee. I am very happy with the ease of loading the bike, securing the items, and the speed of each part as well as the speed overall.

I took minor roads on the way home that were free of the stress and dangers of higher speed, crowded major roads. Now one of my favorite things to do on the way home from an errand as I ride through the neighborhood is to "happen upon" a garage sale. It's amazing how my bike knows which way to turn to follow the garage sales signs! It's kind of funny, but I don't even have to get off the bike to ask "what is the price for that?" I seem to get better prices when I bike in versus arrive via my clean Suburban.

Most importantly, I think all of us e-bikers should start logging our miles so that we can keep track of the dollars that we are saving, build motivation within our e-bike community, and even calculate the pounds of carbon emissions saved by putting in your miles ridden into the carbon footprint calculator at www.ebikebook.com . You can and should become an advocate for e-bike technology and lead by example. People you know and even don't know will ask you how many miles have you put on your e-bike. And you will want to have an answer better than, "ummm I guess hundreds?"

Remember this motivational point and that is that the more miles you use your e-bike, then the less you drive, and the more that EVERYBODY wins. Especially you! I hope you enjoy the journey that is e-bike ownership. I also hope that you will want to show off a picture of your custom designed e-bike and conversion kit installation to your e-biking peers at www.ebikebook.com It is my most sincerest desire to build a free community on the Web for e-bikers to share their results, pictures, reviews, ideas, events, feedback on this book, etc... at www.ebikbook.com . I can't wait to see the picture of your new e-bike. The mission of that website is simply to "build a community that promotes electric bicycles as a green transportation opportunity."

I also see a fit for web sites like www.meetup.com that could play a significant role for the e-bike community as well. I feel the majority of e-bike users are also very "connected via the Web" to their friends and communicate often via the internet. If you have never been to www.meetup.com go there now. Then other websites that I love to promote to Web users are www.yelp.com and www.tripadvisor.com . What they all share in common is the fact that they are driven by user generated information and reviews. Just like I hope a big part of www.ebikebook.com and the related Facebook fan Page will be user generated content.

•

Remember to protect the environment, and please properly dispose of your old batteries as they do not go in your regular trash can. Contact your local Sanitation Dept. for instructions on proper disposal. God Bless America and our beautiful planet!!

Epilogue

For many years now the US and the World have recognized that our fossil fuels will not last forever. And that everyone in the world cannot own and use a car for every single local errand. It appears obvious to me that electric bikes are, and will become more so, a very important long term transportation solution. Not only are they a very economical way to do local travel, but they are much better for the air we breathe, and the world climate.

And yes this is a very significant opportunity even in a car loving metropolis like Los Angeles. On 10-10-10, LA held its first cicLAvia, which was a 7.5 mile route straight through downtown LA, and ending in East Hollywood (see www.cicLAvia.org for details or search www.YouTube.com for cicLAvia) In one 5 hour period on a Sunday starting at 10am, over 100,000 cyclists showed up to use the streets that were free of cars... and yes I did notice a number of e-bikes too! I went with my wife and teenagers. We all had a blast!

Therefore I challenge each and every one of you to drive fewer miles in a car, especially for those short urban trips, weather permitting. If not for the global environmental impact and the reduction of toxic emissions... and if not for exercise and the sheer joy of the wind in your face... then do it for the cost savings, remember the 52 cents per mile incremental cost savings. Ride 1,000 miles every year for short urban trips (which is less than 10% of the average annual amount driven by the average U.S. driver) then you will SAVE $520 a year!!! That's like getting a FREE iPad for you and your family, EVERY YEAR.

Along those lines, I met a fellow e-biker at a Subway Sandwich shop by random luck very recently. He rides his e-bike to work often, and has logged over 3,000 miles in the previous 12 months! That's like saving $1,500 every year!!! He also lost 20 pounds in that one year, and is doing something significant to lessen his personal "carbon pollution footprint" on our precious global environment!!

Thank you very much for reading this book. I sincerely hope you found some good information in it, and it was a good read. If you are an Amazon customer of any product purchase, and feel this book is worth a 4 or hopefully a 5 star rating, please do me the huge favor and write a review of a few sentences on Amazon. I will read every review, and greatly appreciate the rating and help.

I also hope you become an e-bike advocate and lead by example as our car culture ever so slowly coverts. It has to. Don't forget to post pictures of your new e-bike at www.ebikebook.com . I am very excited about the potential of this website, and the potential to create an e-bike community. Please share your experiences related to e-biking and to what is working well for you or not.

Happy Trails, enjoy the journey, and hope to see your bike photos at www.ebikebook.com !!!

Next Steps

Make a final list of what design you want to build focusing on one type "commuter", local errand", "folding bike", or "recreational". Also what bike do you plan to use and most importantly summarize what power, battery type, and conversion kit features are *nice to have,* and what are *must haves*. Before searching the web for conversion kits, go out to my web site www.ebikebook.com to learn from other e-bike users.

Good luck on installing your first e-bike kit, and remember that at some point even a bad decision is better than no decision. In other words, with so many options and issues, it is easy to fall in the trap of "paralysis by analysis." At today's price points, it's time to just "GO FOR IT" and make a decision on a kit. What have you got to lose? $400? The education and experience you get from installing and using your first kit should easily be worth the price of the kit alone!

References

Automotive Exhaust Chemicals: disease causing effects. (n.d.). Retrieved Nov 14, 2010, from Nutramed.com: http://www.nutramed.com/environment/carschemicals.htm *Mitigating Global Warming.*

Weiss, E. (. (2010). *Bike Snob Systematically & Mercilessly Realigning the World of Cycling.* San Francisco: Chronicle Books. LLC.

Slinn, M. (2010). *Build your own Electric Bicycle.* Mc Graw Hill. Smeaton, H. (June, 22 2009).

FBI. (2007). *Crime in the US 2006 (agencies reporting on 24% of US Population).* Retrieved Sept. 19, 2010, from FBI: http://www.fbi.gov/ucr/cius2006/about/index.html

Folding Bicycle Definition ("Mixed-Mode Commuting"). (n.d.). Retrieved Sept. 17, 2010, from Wikipedia: http://en.wikipedia.org/wiki/Folding_bicycle

Insurance Institute for Highway Safety. (n.d.). Retrieved October 17, 2010, from Highway Loss Data Institute: http://www.iihs.org/research/fatality_facts_2008/bicycles.html

Mayo Foundation, f. M. (2010). *The Mayo Clinic Diet: Eat Well, Enjoy Life, Lose Weight.* Good Books.

National Highway Traffic Safety Administration. (n.d.). Retrieved October 17, 2010, from US Dept. of Transportation: http://www-fars.nhtsa.dot.gov/Main/index.aspx

Optibike . (n.d.). Retrieved 12 5, 2010, from The Ferrari of Electric Bikes: www.optibike.com

T., D. (2007, June 14). *Kryptonite's Blog*. Retrieved Sept. 19, 2010, from Unbrreakable Bonds: http://unbreakable-bonds.blogspot.com/2007/06/top-10-cities-for-bike-theft-according.html

Tuff, S. (2009, 6 24). *Gear test - Folding Bikes*. Retrieved 9 17, 2010, from New York Times - Style: http://www.nytimes.com/slideshow/2009/06/24/style/200906 25-physical-slideshow index.html?ref=fashion

Winton, R. (2010, Feb. 11). *LA Now*. Retrieved Sept. 19, 2010, from LA Times - Local Section: http://latimesblogs.latimes.com/lanow/2010/02/la-sees-big-jump-in-bike-thefts-prompting-some-vigilante-justice.html

Why ride electric bikes? Retrieved November 14, 2010, from Suite 101 - Insighful writers: http://www.suite101.com/content/why-ride-an-electric-bicycle-a126792

Index

I hope you find this index helpful. If there are key words that you would like to see indexed that did not make the list, then I would greatly appreciate an email with the key words you would like to see in the index for the next release of this book. As always any feedback or questions are welcome. Please email me anytime at crosay@ebikebook.com

G

H

I

K

L

M

N

O

P

R

S

U

W

Z

CPSIA information can be obtained at www.ICGtesting.com
Printed in the USA
236687LV00007B/143/P